FOCUS ON CLINICAL NEUROPHYSIOLOGY

NEUROLOGY SELF-ASSESSMENT

FOCUS ON CLINICAL NEUROPHYSIOLOGY

NEUROLOGY SELF-ASSESSMENT

Nabil J. Azar, MD
Assistant Professor of Neurology
Associate Director
Clinical Neurophysiology Training Program
Vanderbilt University Medical Center
Nashville, Tennessee

Amir M. Arain, MD
Assistant Professor of Neurology
Director
EEG laboratory
Vanderbilt University Medical Center
Nashville, Tennessee

Wolters Kluwer | Lippincott Williams & Wilkins
Health

Philadelphia · Baltimore · New York · London
Buenos Aires · Hong Kong · Sydney · Tokyo

Acquisitions Editor: Frances R. DeStefano
Managing Editor: Leanne McMillan
Project Manager: Jennifer Harper
Senior Manufacturing Manager: Benjamin Rivera
Marketing Manager: Brian Freiland
Design Coordinator: Holly Reid McLaughlin
Production Services: Laserwords Private Limited, Chennai, India

© 2009 by LIPPINCOTT WILLIAMS & WILKINS, a Wolters Kluwer business

530 Walnut Street
Philadelphia, Pennsylvania 19106 USA
LWW.com

All rights reserved. This book is protected by copyright. No part of this book may be reproduced in any form or by any means, including photocopying, or utilized by any information storage and retrieval system without written permission from the copyright owner, except for brief quotations embodied in critical articles and reviews. Materials appearing in this book prepared by individuals as part of their official duties as U.S. government employees are not covered by the above-mentioned copyright.

Printed in China

Library of Congress Cataloging-in-Publication Data

Azar, Nabil J.
  Focus on clinical neurophysiology / Nabil J. Azar, Amir M. Arain.
    p. ; cm.—(Neurology self-assessment)
  Includes bibliographical references and index.
  ISBN-13: 978-1-58255-854-7 (alk. paper)
  ISBN-10: 1-58255-854-X (alk. paper)
 1. Nervous system—Diseases—Examinations, questions, etc.
2. Neurophysiology—Examinations, questions, etc. I. Arain, Amir M. II. Title.
III. Series.
  [DNLM: 1. Nervous System Physiology—Examination Questions.
2. Diagnostic Techniques, Neurological—Examination Questions.
3. Electrodiagnosis—methods—Examination Questions.
4. Neurophysiology—methods—Examination Questions. WL 18.2 A992f 2009]
  RC343.5.A93 2009
  616.80076—dc22

2008049556

Care has been taken to confirm the accuracy of the information presented and to describe generally accepted practices. However, the authors, editors, and publisher are not responsible for errors or omissions or for any consequences from application of the information in this book and make no warranty, expressed or implied, with respect to the currency, completeness, or accuracy of the contents of the publication. Application of this information in a particular situation remains the professional responsibility of the practitioner.

The authors, editors, and publisher have exerted every effort to ensure that drug selection and dosage set forth in this text are in accordance with current recommendations and practice at the time of publication. However, in view of ongoing research, changes in government regulations, and the constant flow of information relating to drug therapy and drug reactions, the reader is urged to check the package insert for each drug for any change in indications and dosage and for added warnings and precautions. This is particularly important when the recommended agent is a new or infrequently employed drug.

Some drugs and medical devices presented in this publication have Food and Drug Administration (FDA) clearance for limited use in restricted research settings. It is the responsibility of health care providers to ascertain the FDA status of each drug or device planned for use in their clinical practice.

To purchase additional copies of this book, call our customer service department at (800) 638-3030 or fax orders to (301) 223-2320. International customers should call (301) 223-2300.

Visit Lippincott Williams & Wilkins on the Internet: at LWW.com. Lippincott Williams & Wilkins customer service representatives are available from 8:30 AM to 6 PM, EST.

10 9 8 7 6 5 4 3 2 1

PREFACE

There are several very good textbooks available on clinical neurophysiology; however, there is a paucity of study material on technology, methods, and clinical practice to test one's understanding and knowledge. The idea for the book came when we were preparing for the clinical neurophysiology boards. We were surprised by the lack of review books in a question-and-answer format to help us assess our progress and identify areas of weakness needing more attention. We could not evaluate our grasp of the subject. We thought of making notes of the different aspects of clinical neurophysiology that can serve as a review source and evaluation tool.

Moreover, by teaching our residents and fellows, we have realized that they learn better if we challenge them with questions on the subjects. These residents and fellows take their board exams and are tuned to evaluation and assessment by question-and-answer format. This format simulates the board exams and is a great tool to identify areas of strength and weakness. This method also stimulates the reader to read about the areas of weakness and sharpen his acquired knowledge. Similarly people who are practicing clinical neurophysiology as physicians and technologists also like to assess their grasp of the subject and improve on the areas where they have some weaknesses.

The content of this book has been divided into eight main sections: basic electronics and technology, electroencephalography, evoked potentials, nerve conduction studies and electromyography, sleep medicine, autonomic testing, and intraoperative monitoring. This book offers a comprehensive coverage of the subject, supplemented by illustrations. The answers point to and explain the core concepts being tested in the question. This book in no way claims to substitute the textbooks on the subject; rather it, supplements them. We suggest that readers review the answers as they, in simple words, try to explain the concept in its broader aspect. References are given with each answer to further study the concept in detail.

We are very grateful to our current colleagues Dr. Bassel Abou-Khalil and Dr. Peter Donofrio who reviewed sections of this manuscript and provided us with very useful suggestions to improve this book.

We are also indebted to our spouses Joelle and Aneeqa, our children Michel and John, Nidal and Jinan, as well as families and friends who were patient enough to let us work on this book.

Nabil J. Azar, MD, and Amir M. Arain, MD

CONTENTS

SECTION 1: PHYSICS, BASIC PHYSIOLOGY AND TECHNOLOGY

CHAPTER 1 ■	Basic Physics and Electrical Safety	3
CHAPTER 2 ■	Physiological Generators and Volume Conduction	9
CHAPTER 3 ■	Digitization and Instrumentation	15

SECTION 2: ELECTROENCEPHALOGRAPHY

CHAPTER 4 ■	General Principles of Electroencephalography	25
CHAPTER 5 ■	Neonatal and Pediatric Electroencephalography	43
CHAPTER 6 ■	Adult Electroencephalography	61
CHAPTER 7 ■	Clinical Epilepsy	77

SECTION 3: EVOKED POTENTIALS

CHAPTER 8 ■	Visual Evoked Potentials	99
CHAPTER 9 ■	Brainstem Auditory Evoked Potentials	103
CHAPTER 10 ■	Somatosensory Evoked Potentials	109
CHAPTER 11 ■	Evoked Potentials in Clinical Practice	117

SECTION 4: NERVE CONDUCTION STUDIES

CHAPTER 12	■	General Principles of Nerve Conduction Studies	127
CHAPTER 13	■	Sensory Nerve Conduction Studies	133
CHAPTER 14	■	Motor Nerve Conduction Studies	143
CHAPTER 15	■	Nerve Conduction Findings in Common Neuromuscular Disorders	159

SECTION 5: ELECTROMYOGRAPHY

CHAPTER 16	■	General Principles of Electromyography	175
CHAPTER 17	■	Radiculopathies and Motor Neuron Disease	185
CHAPTER 18	■	Plexopathies and Myopathies	195
CHAPTER 19	■	Electromyographic Findings in Common Neuromuscular Disorders	203

SECTION 6: SLEEP MEDICINE

| CHAPTER 20 | ■ | General Principles of Sleep Medicine | 211 |
| CHAPTER 21 | ■ | Sleep Studies in Clinical Practice | 219 |

SECTION 7: AUTONOMIC TESTING AND CENTRAL NEUROPHYSIOLOGY

| CHAPTER 22 | ■ | Autonomic Testing and Central Neurophysiology | 229 |

SECTION 8: NEUROPHYSIOLOGICAL INTRAOPERATIVE MONITORING

| CHAPTER 23 | ■ | Neurophysiologic Intraoperative Monitoring | 239 |

FOCUS ON CLINICAL NEUROPHYSIOLOGY

NEUROLOGY SELF-ASSESSMENT

SECTION 1

Physics, Basic Physiology and Technology

CHAPTER 1

Basic Physics and Electrical Safety

QUESTIONS

1. **A 10-F capacitor is added by a Y-jumper to an already present 10-F capacitor. The new total capacitance is:**
   A. 0 F
   B. 5 F
   C. 10 F
   D. 20 F

2. **The threshold for induction of ventricular fibrillation is approximately:**
   A. 10 A
   B. 1 A
   C. 100 mA
   D. 10 mA

3. **Kirchhoff's laws define the principle of:**
   A. Dissipation of energy
   B. Conservation of energy
   C. Dissipation of resistances
   D. Conservation of resistance

4. **Which of the following is true about a capacitor?**
   A. Consists of two charged conductors separated by a resistor
   B. Is a device that stores a magnetic field
   C. Its charge q is inversely proportional to the potential difference
   D. Has a capacitance measured in Farad (F)

5. An inductor (coil) is defined by all of the following except:
   A. Generates a magnetic field when traversed by a current
   B. Has a property called *inductance*
   C. Is measured in seconds per ampere or Henry
   D. Facilitates a change in current flow

6. Electric shock is defined by all of the following except:
   A. Port of entry and exit
   B. Current source and ground, both connected to the body
   C. High resistance path
   D. Flow of current through the body

7. What is the maximal allowed leakage current in an intensive care unit (ICU) patient with a central venous pressure catheter?
   A. 100 mA
   B. 20 mA
   C. 100 μA
   D. 20 μA

8. A Y-jumper with a resistance of 10,000 Ω is added to an already present resistor of 10,000 Ω, the new total resistance is:
   A. 20,000 Ω
   B. 10,000 Ω
   C. 5,000 Ω
   D. 0 Ω

9. A requirement for a lethal shock is that the current pass through the:
   A. Brain
   B. Heart
   C. Chest
   D. All of the above

10. The voltage (V) is related to the current (I) and the resistance (R) by the equation:
    A. $V = I/R$
    B. $V = R/I$
    C. $V = I \times R$
    D. $V = 1/I + 1/R$

11. Reduction of electrocution risk includes all of the following measures except:
    A. Minimizing leakage current
    B. Connecting technologists to the ground
    C. Ensuring high resistance contacts with source of leakage current and with ground
    D. Using one ground per patient

12. In healthy patients, the chassis leakage current should not exceed:
    A. 100 mA
    B. 10 mA

C. 100 μA
D. 10 μA

13. **Methods to reduce leakage current include all except:**
   A. Use of an equipotential grounding system
   B. Use of nonmetallic appliances
   C. Use of appliances with long line cords
   D. Use of current limiters

ANSWERS

1. (D): Similarly to resistors, capacitors can be placed in series or parallel. But because of the differences in their electrical properties, their formulae are reversed. The total capacitance of capacitors in parallel (mounted by Y-jumper) is equal to the sum of all the individual capacitances. The total capacitance of capacitors placed in series is the inverse of the reciprocal of the sum of individual capacitances. (**Misulis and Head 2003, pp. 30–31**)

2. (C): The threshold current for induction of ventricular fibrillation is approximately 100 mA. Because this current spreads out as it passes through the body, only a small fraction will pass through the heart. Consequently, if applied directly to the heart, 0.1 mA may be enough to induce ventricular fibrillation. (**Ebersole and Pedley 2003, pp. 66–67; Daube 2002, p. 19**)

3. (B): Kirchhoff's laws define the principle of conservation of energy. Kirchhoff's current law states that at any point in an electrical circuit where charge density is not changing in time, the sum of currents flowing toward that point is equal to the sum of currents flowing away from that point.

Kirchhoff's voltage law states that the directed sum of the electrical potential differences around any closed circuit must be zero. (**Daube 2002, p. 5; Misulis and Head 2003, pp. 27–28**)

4. (D): A capacitor is a device that stores electric charge. It consists of two charged conductors separated by an insulator (dielectric) such that no charge can pass through but charges segregate on both of its sides. After it is fully charged, it will act as battery when the current source is shut down. Its charge q is proportional to the potential difference V ($q = CV$) where C is the capacitance measured in coulombs per volt or Farad (F). (**Ebersole and Pedley 2003, pp. 35–36; Misulis and Head 2003, p. 19**)

5. (D): An inductor or electromagnet consists of coils of wire that generates a magnetic field when traversed by a current. An inductor has the property of resisting any change in current flow by producing an electromagnetic field. It has a property called *inductance* and is measured in seconds per ampere or Henry (H). Inductance has the same relationship to current that capacitance has to voltage. (**Daube 2002, p. 5**)

6. (C): Electric shock results from current flow through the body. It requires two connections to the body—an entry point, which is the apparatus and an exit point, which is the ground. It also requires a low resistance pathway. (**Daube 2002, pp. 17–18**)

7. (D): Higher risk of electrocution occurs in susceptible patients. Premorbid heart conditions, pacemakers, and invasive instrumentation are all risk factors for ventricular fibrillation. In these patients the maximal allowed leakage current is 10 to 20 μA. (**Ebersole and Pedley 2003, pp. 65–66; Daube 2002, p. 25**)

8. (C): The total resistance across resistors in parallel (mounted by a Y-jumper) is the reciprocal of the sum of the inverse of each resistor. Conversely, the total resistance across resistors mounted in series is the sum of all resistors. (**Misulis and Head 2003, pp. 28–30; Daube 2002, pp. 6–7**)

9. (B): The mechanism of lethal shock is almost always secondary to ventricular fibrillation. Therefore, the electric current should take a path through the body that includes the heart to cause death. Complications of other organ failure, severe burns, and injuries may also result in severe morbidity or even mortality. (**Daube 2002, pp. 18–19**)

10. (C): Ohm's law defines the relationship between the voltage (V), the current (I), and the resistance (R). In a closed circuit, the voltage is equal to the product of the intensity and resistance. (**Misulis and Head 2003, pp. 25–26; Ebersole and Pedley 2003, pp. 34–35**)

11. (B): Several factors can reduce the risk of electrical shock in neurophysiologic testing. Of these are: use of equipments with minimal leakage currents, use of one common ground per patient, use of high resistance contacts with the source and ground such as dry and intact skin. People using appliances should not be connected to the ground. (**Daube 2002, p. 20, Ebersole and Pedley 2003, p. 67**)

12. (C): The total chassis-to-earth ground leakage current from each terminal connected to the patient should not exceed 100 μA. In susceptible patients with known heart conditions, the chassis leakage current should not exceed 20 μA. (**Ebersole and Pedley 2003, p. 67; Daube 2002, p. 25**)

13. (C): Many methods are used in biomedical instruments to reduce leakage current. These include the use of equipotential grounding system, use of nonmetallic appliances, use of current limiters with leakage current of <20 μA. Extension cords or long line cords should never be used as each one foot of line cord adds 1 μA of leakage current into the ground connection. (**Daube 2002, pp. 22–23; Ebersole and Pedley 2003, pp. 67–68**)

References

1. Daube J. *Clinical neurophysiology*, 2nd ed. Oxford Press; 2002.
2. Ebersole John, Pedley Timothy. *Current practice of clinical electroencephalograph*, 3rd ed. Lippincott Williams & Wilkins; 2003.
3. Misulis K, Head T. *Essentials of clinical neurophysiology*, 3rd ed. Butterworth-Heineman; 2003.

CHAPTER 2

Physiological Generators and Volume Conduction

QUESTIONS

1. The resting membrane potential of neurons is approximately:
   A. +70 μV
   B. −70 μV
   C. +70 mV
   D. −70 mV

2. The ion with the highest intracellular concentration is:
   A. Sodium
   B. Potassium
   C. Chloride
   D. Calcium

3. The experimental mathematical model of the action potential was developed by:
   A. Watson and Clark
   B. Kandel and Schwartz
   C. Hodgkin and Huxley
   D. Azar and Arain

4. At equilibrium, the Nernst potential of sodium is:
   A. +60 mV
   B. −60 mV

C. +10 mV
D. −10 mV

5. **Slow calcium spikes are involved in:**
   A. Early afterdepolarization
   B. Intrinsic bursting behavior of pyramidal cells
   C. Paroxysmal depolarization shift
   D. All of the above

6. **Inhibitory postsynaptic potentials (IPSPs) are mostly mediated by the opening of which type of channel?**
   A. Chloride
   B. Chloride and potassium
   C. Sodium and calcium
   D. Sodium

7. **The lowest threshold for neuronal membrane depolarization occurs at the:**
   A. Cell body
   B. Dendrite
   C. Axon hillock
   D. Synapse

8. **The myelin sheath has all of the following properties except:**
   A. Increases the time constant
   B. Increases speed of propagation of action potentials
   C. Decreases membrane capacitance and conductance
   D. Increases the space constant of axons along the nodes of Ranvier

9. **An excitatory postsynaptic potential (EPSP) located superficially in the cerebral cortex will produce a dipole with:**
   A. Superficial positive and deep negative poles
   B. Superficial negative and deep positive poles
   C. Superficial positive and negative poles
   D. Deep positive and negative poles

10. **In the cerebral cortex, the generator of spontaneous electroencephalographic (EEG) activity is defined by all of the following except:**
    A. It is mainly composed of pyramidal cells
    B. Pyramidal cells are arranged with the main axis of the dendritic trees parallel to each other and perpendicular to the cortical surface
    C. The longitudinal component of the current flow from different neurons cancel each other and the transverse component add together
    D. Thousands of pyramidal cells are usually simultaneously activated by a single axon

11. **The neocortex consists of:**
    A. Three cellular layers
    B. Four cellular layers
    C. Five cellular layers
    D. Six cellular layers

12. **In the resting state, all of the following net ion movements are correct except:**
    A. Net flux of sodium into the cell
    B. Net flux of potassium outside the cell
    C. Net flux of chloride into the cell
    D. All of the above

13. **The thalamus is the generator of:**
    A. Spikes
    B. Posterior dominant rhythm
    C. Sleep spindles
    D. All of the above

14. **Afterhyperpolarizations are primarily mediated by:**
    A. Calcium-dependent sodium channels
    B. Calcium-dependent potassium channels
    C. Chloride-dependent sodium channels
    D. Chloride-dependent potassium channels

ANSWERS

1. **(D):** The resting membrane potential of neuron and muscle fibers is approximately −70 mV. This potential is defined as the electric potential inside the cell minus that outside the cell. This potential is maintained as a result of ionic concentration gradient across the semipermeable cell membrane. **(Daube 2002, p. 53)**

2. **(B):** The value of the resting membrane potential depends on the concentration of ions across the semipermeable cell membrane. The ion with the highest intracellular concentration is potassium (68 mmol/L) and the ions with the highest extracellular concentrations are sodium (142 mmol/L) followed by chloride (105 mmol/L). **(Daube 2002, p. 54)**

3. **(C):** The Hodgkin-Huxley model is a scientific model that describes how action potentials in neurons are initiated and propagated. This mathematical model was based on experimental measurements of the time and voltage dependence of sodium and potassium conductances in giant squid axons. Drs. Hodgkin and Huxley received the 1963 Nobel Prize in Physiology for this work. **(Daube 2002, p. 56)**

4. **(A):** The Nernst potential or electromotive force of every ion is determined by its ionic concentration, its conductance per unit area and the capacitance of the membrane. In the steady state, sodium has a Nernst potential of +60 mV and has the highest extracellular concentration. Potassium has a Nernst potential of −66 mV and has the highest intracellular concentration. **(Daube 2002, pp. 53–54)**

5. (D): Slow calcium spikes are depolarizing potentials mediated by slow calcium channels that are activated by membrane depolarization and inactivated by repolarization and increased intracellular calcium concentrations. They are involved in the intrinsic bursting behavior of pyramidal cells and the formation of the paroxysmal depolarization shift (PDS). They are also involved in early afterdepolarization. (**Ebersole and Pedley 2003, pp. 4–5; Daube 2002, pp. 58–59**)

6. (B): Postsynaptic potentials are nonpropagating nerve or muscle potentials caused by neurotransmitter-induced opening or closing of channels. These excitatory or inhibitory potentials constitute normal synaptic transmission. EPSPs are mostly mediated by opening of sodium channels allowing entry of sodium inside the cell causing cell depolarization. IPSPs are mostly mediated by opening of chloride channels allowing entry of chloride inside the cell and potassium channels allowing exit of potassium outside the cell, causing cell hyperpolarization. (**Fisch 1999, pp. 6–7; Daube 2002, p. 57**)

7. (C): Action potentials are referred to as all-or-none phenomenon. They occur when the neuronal membrane is depolarized beyond a critical threshold. This threshold is lowest at the axon hillock located at the junction of the axon to the cell body. Action potentials have an amplitude of approximately 110 mV and lasts about 1 ms. (**Fisch 1999, p. 8**)

8. (A): Action potentials velocity is faster in myelinated fibers, compared to unmyelinated ones. The myelin sheath alters the cable properties of axons. The myelin sheath decreases membrane capacitance and conductance, decreases the time constant and increases the space constant along the nodes of Ranvier. (**Daube 2002, p. 55**)

9. (B): An EPSP occurs when positive ions flow intracellularly thereby creating a negative potential. An ESPS located superficially in the cerebral cortex along the distal branches of the pyramidal cells will produce a dipole with a superficial negative and a deep positive pole. (**Daube 2002, p. 29**)

10. (C): Because pyramidal cells are arranged in a regular manner, with main axis of the dendritic trees parallel to each other and perpendicular to the cortical surface, the longitudinal component of the current flow from different neurons add together and the transverse component cancel out, producing a laminar current along the main axis of neurons. (**Daube 2002, p. 29**)

11. (D): The neocortex is composed of six cellular layers labelled I to VI (with VI being the innermost and I being the outermost). The neocortex contains two primary types of neurons, excitatory pyramidal neurons (~80% of neocortical neurons) and inhibitory interneurons (~20%). Its structure is relatively uniform with six horizontal layers segregated principally by cell type and neuronal connections. (**Daube 2002, p. 61**)

12. (C): The net ionic flux across the membrane is the product of the net driving force and the membrane permeability of an ion. In the resting potential, there is flux of ions down their potential gradient. There is net flux of sodium into the cell, potassium outside the cell and no net flux of chloride. The steady state is maintained by the sodium/potassium pump. (**Daube 2002, p. 54**)

13. (C): Sleep spindles are generated in the thalamus. Their morphology and degree of synchronization is influenced by the cerebral cortex. This was demonstrated in animals models where decortication did not abolish sleep spindles. (**Niedermeyer and Lopes Da Silva 1999, pp. 29–30**)

14. (B): Afterhyperpolarizations are prolonged hyperpolarizing potentials that limit the firing rate of cells following sodium or calcium mediated depolarizing potentials. They are primarily mediated by calcium-dependent potassium channels. These channels are activated by increased intracellular calcium concentrations and membrane depolarization whereas decreased intracellular calcium concentration and membrane repolarization inhibit them. (**Daube 2002, p. 58**)

References

1. Daube J. *Clinical neurophysiology*, 2nd ed. Oxford Press; 2002.
2. Ebersole J, Pedley T. *Current practice of clinical electroencephalograph*, 3rd ed. Lippincott Williams & Wilkins; 2003.
3. Fisch B. *Fisch and and Spehlmann's EEG primer: principles of digital and analog EEG*, 3rd ed. Elsevier Science; 1999.
4. Niedermeyer Ernst, Lopes Da Silva F. *Electroencephalography: basic principles, clinical applications, and related fields*, 4th ed. Lippincott Williams & Wilkins; 1999.

CHAPTER 3

Digitization and Instrumentation

QUESTIONS

1. **Which of the following properties apply to amplifiers used in clinical neurophysiology?**
   A. Differential amplifier
   B. It amplifies all potentials recorded at every input
   C. Common rejection mode of 1,000
   D. All of the above

2. **What is the minimal recommended resolution power of an analog-to-digital converter?**
   A. 8 bits
   B. 10 bits
   C. 12 bits
   D. 14 bits

3. **Compared to analog recording, digital electroencephalography (EEG) recording has the advantage of:**
   A. Better inter-reader agreement
   B. Lower cost
   C. Standardized data formats
   D. Limited information technology (IT) maintenance

4. **A computer monitor has a resolution of 1,024 × 768 pixels. On each side, 50 pixels are devoted for screen border. If 10 seconds of EEG are displayed on the screen, how many pixels are devoted to display 1 second of horizontal EEG?**
   A. 102 pixels
   B. 92 pixels

C. 66 pixels
D. 56 pixels

5. **In electroencephalography, electrode impedance should be between:**
   A. 1 to 10 Ω
   B. 10 to 100 Ω
   C. 100 to 5,000 Ω
   D. 5,000 to 10,000 Ω

6. **Aliasing refers to:**
   A. Misinterpretation of amplitudes throughout digitization process
   B. Distortion of a signal caused by sampling frequency lower than the Nyquist frequency
   C. Prolongation of the waveform latencies
   D. 60-Hz artifact

7. **The ability of an amplifier to reject in phase and amplify out of phase potentials is known as:**
   A. Amplifier gain
   B. Aliasing
   C. Common mode rejection
   D. Analog to digital conversion

8. **A high-pass filter develops its output potential across a:**
   A. Capacitor
   B. Resistor
   C. Inductor
   D. Amplifier

9. **Well-ventilated areas are required when using:**
   A. Paraffin wax
   B. Collodion
   C. Isopropyl alcohol
   D. Patex

10. **Which of the following is most likely to happen if electrode impedance is measured to be 50 Ω?**
    A. Stable waveform recording
    B. Short circuiting of EEG potential differences
    C. Longer time constant
    D. Phase shift

11. **What is the lowest sampling frequency needed to resolve a 60-Hz signal?**
    A. 30 Hz
    B. 60 Hz
    C. 120 Hz
    D. 240 Hz

12. **The resolution of digital signals is better than analog signals.**
    A. True
    B. False

Chapter 3: Digitization and Instrumentation

13. **Analog to digital conversion requires:**
    A. Digitization
    B. Quantization
    C. Sampling
    D. All of the above

14. **A low-pass filter develops its output potential across a:**
    A. Capacitor
    B. Resistor
    C. Inductor
    D. Amplifier

15. **A high-pass filter causes the square-wave calibration curve to:**
    A. Diminish exponentially
    B. Augment exponentially
    C. Reverse polarity
    D. Does not cause any changes

16. **The time constant TC of low-frequency filters is:**
    A. Directly proportional to the *cut-off* filter frequency
    B. Inversely proportional to the *cut-off* filter frequency
    C. Equivalent to the *cut-off* filter frequency
    D. Not related to the *cut-off* filter frequency

17. **To double the signal-to-noise ratio, the number of sweeps must increase by a factor of:**
    A. 2
    B. 4
    C. 8
    D. 16

18. **To avoid aliasing, sampling rate must be:**
    A. At least twice that of the lowest sampled frequency
    B. At least twice that of the highest sampled frequency
    C. At least half that of the lowest sampled frequency
    D. At least half that of the highest sampled frequency

19. **To limit the drop in signal amplitude to 1%, the amplifier impedance should:**
    A. Match the electrode impedance
    B. Be at least ten times less than electrode impedance
    C. Be at least 100 times higher than electrode impedance
    D. Not be compared to electrode impedance

20. **In electromyography, compared to monopolar needle electrodes, concentric needle electrodes will cause the amplitude to be:**
    A. Smaller
    B. Larger
    C. Same
    D. Either smaller or larger

21. **A higher display sensitivity or gain results in:**
    A. Longer onset latency
    B. Shorter onset latency
    C. Higher signal amplitude
    D. Lower signal amplitude

22. **Which of the following neurophysiologic signals is expressed in mV?**
    A. Somatosensory evoked potential (SSEP)
    B. Compound motor action potential (CMAP)
    C. Sensory nerve action potential (SNAP)
    D. All of the above

23. **60-Hz artifact may be caused by:**
    A. Electrode impedance mismatch
    B. Broken electrode
    C. Loosely attached electrodes
    D. All of the above

ANSWERS

1. (A): The differential amplifier used in clinical neurophysiology amplifies the differences in potentials between its two inputs. This allows amplification of potential differences and cancellation of contaminating potentials present at both inputs like in the case of 60-Hz noise from line voltage devices. (**Daube 2002, pp. 15–16**)

2. (C): Digital signals are quantified in discrete nonoverlapping amplitude levels, expressed in terms of bits where every bit is a power of two.

Signal amplitude of analog-to-digital converter (ADC) depends on both its voltage range and the maximum number of bits discrimination it can resolve.

ADCs used in both EEG and evoked potential recordings should have a minimum of 12 bits ($12^2 = 4,096$ voltage levels spread over $+2,048$ and $-2,048$). This will allow good visual resolution of signals $<1\mu V$. (**Fisch 1999, pp. 60–62**)

3. (A): Digital recording allows linear display, convenient storage and retrieval, montage reformatting, filter, sensitivity and time-based changes, and annotation recording. All these features have been shown to enhance reliability of interpretation. The disadvantages of digital recording when compared to analog recording are cost, incompatible data formats and softwares, and need for close maintenance by IT services. (**Daube 2002, pp. 42–43**)

4. (B): In digital EEG display, monitor resolution is the number of pixels in each direction that can be used to fill in points on the EEG tracing.

In this case, of the 1,024 horizontal pixels, 100 pixels are higher used for screen margin leaving 924 pixels for a 10 second of EEG. One second of EEG display will have approximately 92 pixels devoted to it.

5. (C): Before EEG recording, the impedance of each electrode should be routinely measured. This is performed with an impedance meter to evaluate the contact of each electrode with the scalp. The impedance should be between 100 and 5,000 Ω. **(Fisch 1999, p. 25)**

6. (B): The Sampling Theorem states that if a signal contains component frequencies ranging from 0 to fN, then the minimum sampling frequency that can be used for a digitized data to adequately represent the frequency content of the original signal is 2 fN, called the Nyquist frequency (equal to two times of the original frequency sampled). Aliasing refers to the distortion of a signal caused by sampling frequency lower than the Nyquist frequency. **(Ebersole and Pedley 2003, p. 47; Daube 2002, pp. 46–47)**

7. (C): EEG amplifiers are designed to amplify out of phase potentials to reveal differences in electrical potentials between its two inputs and to reject potentials that are common to the two inputs. This property is referred to as *common mode rejection*. **(Fisch 1999, pp. 40–42)**

8. (B): A high-pass filter (low-frequency filter) consists of a capacitor in series and a resistor in parallel. Its output current potential is generated across a resistor. Filters are used for noise attenuation and identification of signals, immeasurable under normal recording conditions. On the other hand, filters distort waveforms because of their frequency attenuation and phase shift characteristics. **(Daube 2002, pp. 12–13)**

9. (B): Good mechanical and chemical contact of electrodes to the scalp is achieved with the use of collodion. Because of the chemical interactions with acetone (a substance used to remove collodion), a well-ventilated area is required for safe use of these chemicals. **(Fisch 1999, pp. 20–21)**

10. (B): Electrode impedance evaluates the contact between the electrode and the scalp. Electrode impedance should measure >100 Ω and <5,000 Ω for stable waveform recording. Very low impedance will act like a shunt between the recording electrodes and will short circuit EEG potential differences.

An electrode with very low impedance often makes contact with other electrode as a result of excess electrode paste or sweat connecting the two electrodes. **(Fisch 1999, pp. 25–26)**

11. (C): To faithfully represent a particular frequency, the sampling rate should be at least equal to the Nyquist frequency that is twice the frequency of the signal resolved. If sampled at a lower rate, distortion of the waveform occurs, a phenomenon referred to as *aliasing*. **(Ebersole and Pedley 2003, p. 47; Fisch 1999, pp. 58–59)**

12. (B): EEG records continuous brain activity that is filtered and amplified. It is continuous and uninterrupted and as such remains an analog signal, analogous to the source signal. When digitized, it becomes discontinuous and less smooth because of sampling. **(Ebersole and Pedley 2003, pp. 44–45)**

13. (D): Digitization is the process by which analog signals are converted to digital signals. Quantization consists of assigning a value to every potential input to the analog-to-digital converter. Sampling determines the temporal resolution and the frequency needed for waveform reproduction. (**Ebersole and Pedley 2003, pp. 44–45; Daube 2002, pp. 44–45**)

14. (A): A low-pass filter (high frequency filter) consists of a resistor in series and a capacitor in parallel. Its output current potential is generated across a capacitor. (**Daube 2002, pp. 12–13**)

15. (A): A high-filter pass causes the output current to attenuate at low frequency and thus will cause the square-wave calibration curve to diminish exponentially to a set frequency cut-off with an attenuation factor of 67%. On the contrary a low-pass filter has the opposite effect, attenuating high frequency output current causing an exponentially rising square-wave calibration curve with a set cut-off frequency and with an attenuation of 33%. (**Daube 2002, pp. 12–13**)

16. (B): The time constant is a measure of the low-frequency filters on square pulses. It is the time required for 63% fall of the calibration square-wave pulse at steady voltage. The cut-off filter frequency f is inversely proportional to the time constant TC and vice versa: $f = 1/(2\pi \text{TC})$.

In other words, if the time constant is increased, the low-frequency filter is lowered which will reduce the filtering of low frequency waves. (**Daube 2002, pp. 12–13; Niedermeyer and Lopes Da Silva 1999, pp. 116–117**)

17. (B): When signal amplitude is very small compared to background noise (such as in sensory nerve action potential [SNAP] and somatosensory evoked potential [SSEP]), waveform averaging is used to increase the signal-to-noise ratio (SNR). SNR is proportional to the square root of the number of averaged trials. Therefore, to double the SNR, four sweeps should be used. (**Chiappa 1997, p. 581; Daube 2002, p. 74**)

18. (B): To avoid distortion of waveforms or aliasing, sampling rate should be at least twice that of the highest frequency sampled. This frequency is referred to as the *Nyquist frequency*. (**Ebersole and Pedley 2003, p. 47**)

19. (C): The built-in electrode impedance to current flow attenuates the voltage that is perceived by the amplifier. To limit this attenuation, the impedance of the amplifier should be much higher than the electrode. To limit the drop in signal amplitude to 1%, the amplifier impedance should be at least 100 times more than the electrode impedance. (**Kimura 2001, pp. 43–44**)

20. (A): Compared to monopolar needle electrodes, concentric needle electrodes have the reference electrode within millimeters from the active electrode. The proximity of the two electrodes may cause reduction of the motor unit action potential (MUAP) resulting in lower amplitudes. For the same reason, it may also reduce or cancel artifact or noise. (**Preston and Shapiro 2005, pp. 163–165**)

21. (B): Sensitivity and sweep speed can both affect the recorded latency of sensory or motor responses. A higher sensitivity will allow the detection of smaller deflections from baseline, causing shorter onset latency and longer duration. This effect is more important in detecting onset latencies than peak latencies. Comparative studies or studies within the same nerve should be recorded at the same sensitivity and sweep speed. (**Preston and Shapiro 2005, p. 103**)

22. (B): The evoked potential response is expressed in few μVs, often $<1\mu$V. The SNAP response is expressed in several μVs. The CMAP and MUAP are expressed in several mVs. Electronic amplification increases the amplitude of the action potentials to recognizable levels, a phenomenon referred to as gain. (**Kimura 2001, pp. 94–95; Chiappa 1997, pp. 26–27**)

23. (D): Electrical noise, such as 60-Hz interference by other devices, is common in electrodiagnostic studies. This noise is more common in the intensive care unit where multiple devices are used. It can greatly influence the reading of small-recorded potentials such as evoked and sensory potentials.

It is commonly caused by electrode impedance mismatch. The result of this mismatch is augmented by the differential amplifier. Other causes are broken or frayed electrodes and loose connections to the recording site. (**Preston and Shapiro 2005, p. 91**)

References

1. Chiappa KH. *Evoked potentials in clinical medicine*, 3rd ed. Lippincott-Raven; 1997.
2. Daube J. *Clinical neurophysiology*, 2nd ed. Oxford Press; 2002.
3. Ebersole J, Pedley T. *Current practice of clinical electroencephalograph*, 3rd ed. Lippincott Williams & Wilkins; 2003.
4. Fisch B. *Fisch and Spehlmann's EEG primer: principles of digital and analog EEG*, 3rd ed. Elsevier Science; 1999.
5. Kimura J. *Electrodiagnosis in diseases of nerve and muscle: principles and practice*, 3rd ed. Oxford University Press; 2001.
6. Niedermeyer E, Lopes Da Silva F. *Electroencephalography: basic principles, clinical applications, and related fields*, 4th ed. Lippincott Williams & Wilkins; 1999.
7. Preston DC, Shapiro BE. *Electromyography and neuromuscular disorders: clinical-electrophysiologic correlations*, 2nd ed. Butterworth-Heineman; 2005.

SECTION 2

Electroencephalography

CHAPTER 4

General Principles of Electroencephalography

QUESTIONS

1. Which of the following electrodes is nonpolarizable?
   A. Gold
   B. Platinum
   C. Silver
   D. Silver-silver chloride

2. Using the international 10-20 electrode system, if the distance between the nasion and the inion is 38 cm, what would the distance from Fz to Pz be?
   A. 15.2 cm
   B. 19.0 cm
   C. 27.8 cm
   D. 30.2 cm

3. The following is true about the combinatorial nomenclature of the 10-10 system as compared to the international 10-20 system:
   A. T7/T8 replace T3/T4
   B. P7/P8 replace T5/T6
   C. A and B
   D. None of the above, there is no difference in electrode naming

4. High-frequency filter of 30 Hz would most attenuate:
   A. 15 Hz
   B. 30 Hz

C. 55 Hz
D. 70 Hz

5. **Amplitude of scalp-recorded electroencephalogram (EEG) potentials depends on:**
   A. Intensity of the potential
   B. Spatial orientation of the potential
   C. Resistance of the structures
   D. All of the above

6. **Compared to near-field potentials, far-field potentials typically have:**
   A. Smaller scalp distribution
   B. Wider scalp distribution
   C. Higher voltage
   D. Better localizing value

7. **Increasing the number of seconds displayed on the monitor:**
   A. Decreases spike amplitude
   B. Decreases spike duration
   C. Decreases spike frequency
   D. Does not change spike amplitude, duration, or frequency

8. **Which of the following best describes the amplitude of the scalp potential of an EEG signal?**
   A. It is proportional to the solid angle of the cortical generator at the electrode site
   B. It is inversely proportional to the solid angle of the cortical generator at the electrode site
   C. It is proportional to the distance from the cortical generator to the recording electrode
   D. None of the above

9. **The intracellular pathophysiologic mechanism of a focal epileptiform discharge is:**
   A. Action potential
   B. Paroxysmal depolarizing shift
   C. Presynaptic excitation
   D. Hyperpolarization

10. **The ipsilateral ear montage shows prominent electrocardiogram (ECG) artifact greater on the left. Which of the following is the best strategy to minimize the artifact?**
    A. Switch to contralateral ear reference
    B. Switch to left ear reference
    C. Switch to right ear reference
    D. Using a linked ear reference

11. **In a bipolar montage, all sharp waves will be displayed as a reversal of polarity.**
    A. True
    B. False

Chapter 4: General Principles of Electroencephalography   27

12. **Which thalamic nucleus does not project directly to the cortex?**
    A. Pulvinar
    B. Reticular nucleus
    C. Anterior nucleus
    D. By definition, all thalamic nuclei project directly to the cortex

13. **Which of the following electrode pairs are most sensitive respectively to vertical and horizontal eye movements?**
    A. T3/T4 and F7/F8
    B. Fp1/Fp2 and F7/F8
    C. F3/F4 and T3/T4
    D. Fp1/Fp2 and T3/T4

14. **During long-term video-EEG recording, which of the following material is preferably used to attach electrodes to the scalp?**
    A. Acetone
    B. Collodion
    C. Gorilla glue
    D. Super glue

15. **Which of the following dipoles contribute least to the scalp EEG?**
    A. Horizontal
    B. Radial
    C. Circular
    D. Oblique

16. **The maximal voltage of a sharply contoured activity at F3 is $-70\ \mu V$ and at Fz is $-30\ \mu V$. What would the voltage be across the channel Fz-F3?**
    A. $-40\ \mu V$
    B. $40\ \mu V$
    C. $-100\ \mu V$
    D. $100\ \mu V$

17. **γ Aminobutyric acid A (GABA$_A$) receptors modulate:**
    A. Sodium channels
    B. Potassium channels
    C. Chloride channels
    D. Calcium channels

18. **What are the electrical charges of the cornea and tip of the tongue?**
    A. Cornea and tip of tongue are positively charged
    B. Cornea and tip of tongue are negatively charged
    C. Cornea is positively charged and tip of tongue is negatively charged
    D. Cornea is negatively charged and tip of tongue is positively charged

19. **The duration range of a sharp wave is:**
    A. 0–70 ms
    B. 70–200 ms
    C. 200–370 ms
    D. It is not defined by duration

20. In the 10-20 international nomenclature, EEG activity from the F7 and F8 electrodes localizes to the:
    A. Inferior frontal region
    B. Anterior temporal region
    C. Mesial frontal region
    D. A or B

21. In the intensive care unit (ICU), a ground loop can be avoided by:
    A. Using multiple patient grounds
    B. Using a long extension cord
    C. Connecting each piece of equipment to a separate outlet ground
    D. Using one common ground

22. Tetrodotoxin is a:
    A. Sodium channel blocker
    B. Calcium channel blocker
    C. Potassium channel blocker
    D. GABA receptor blocker

23. What is the mechanism of action of baclofen, a drug than lowers seizure threshold?
    A. $GABA_A$ agonist
    B. $GABA_A$ antagonist
    C. $GABA_B$ agonist
    D. $GABA_B$ antagonist

24. Scalp-recorded electrical activity arises from:
    A. Action potentials generated in the deep layers of the cortex
    B. Action potentials generated in the superficial layers of the cortex
    C. Postsynaptic potentials generated in the superficial layers of the cortex
    D. Postsynaptic potentials generated in the deep layers of the cortex

25. In order to record a spike on scalp EEG, the minimal surface area of cortex involved is:
    A. 0.6 mm^2
    B. 6 mm^2
    C. 0.6 cm^2
    D. 6 cm^2

26. The EEG recording shown in Figure 4.1 is most consistent with:
    A. Roving eye movements
    B. Ictal nystagmus
    C. Eye flutter
    D. Breach rhythm

27. Bancaud phenomenon is a normal EEG variant.
    A. True
    B. False

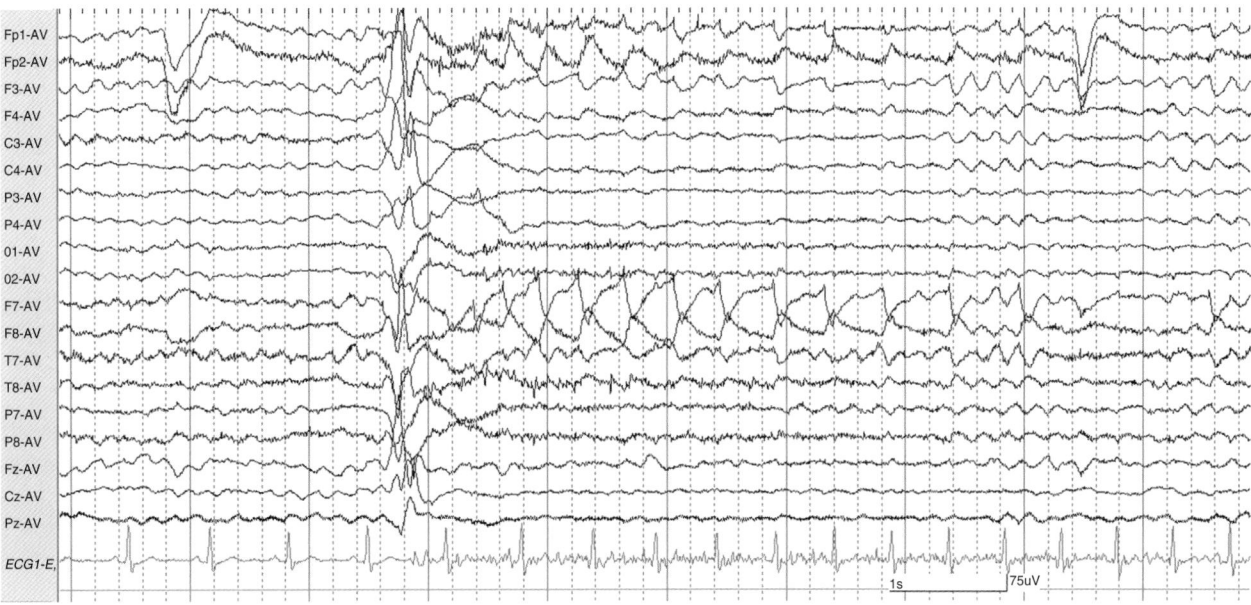

Figure 4.1.

28. **Sphenoidal and zygomatic electrodes record brain activity, respectively, from:**
    A. Inferior-mesial and superior-lateral temporal regions
    B. Superior-lateral temporal and inferior-mesial temporal regions
    C. Inferior-mesial and inferior-lateral temporal regions
    D. Both record from inferior-mesial temporal regions

29. **When filtered, the glossokinetic artifact will look like:**
    A. Rhythmic delta activity
    B. Beta activity
    C. Repetitive sharp transients
    D. Rhythmic alpha activity

30. **All of the following terminology is true except:**
    A. When applied to a single wave, the frequency in Hz is obtained by: 1/wavelength (seconds)
    B. Arrhythmic activity is irregular
    C. Rhythmic activity is the same as periodic activity
    D. A complex is composed of a combination of two or more waves such as a spike-and-wave

31. **The pathophysiology of focal slow waves is due to:**
    A. Cortical hyperexcitability
    B. Partial deafferentation of cortex from subcortical structures
    C. Thalamic synchronization
    D. Paroxysmal depolarizing shift

32. **In routine adult EEG recording, all of the following are considered standard settings except:**
    A. Low-frequency filter of 1 Hz and high-frequency filter of 70 Hz
    B. Sensitivity of 7 μV per mm

C. Paper speed of 10 mm per second
D. Length of recording of 20 minutes

33. **Which of the following measures may help enhance focal slow activity?**
    A. Decrease the speed of recording, decrease the time constant and decrease the sensitivity
    B. Decrease the speed of recording, increase the time constant and increase sensitivity
    C. Increase the speed of recording, decrease the time constant and increase the sensitivity
    D. Increase the speed of recording, increase the time constant and decrease the sensitivity

34. **Which of the following montages displays best temporal lobe activity during wakefulness?**
    A. Linked ears referential montage
    B. Ipsilateral ear referential montage
    C. Vertex referential montage
    D. Average referential montage

35. **Which of the following is true about response to photic stimulation?**
    A. Most prominent in newborn
    B. Higher amplitude response is obtained with eyes closed
    C. Largest response is seen at frequencies >20 Hz
    D. Best seen in the parietal electrodes

36. **Which of the following features may help distinguish epileptiform from nonepileptiform sharp transients?**
    A. Polyphasic morphology
    B. High amplitude
    C. After-going slow wave
    D. All of the above

37. **The potential that demonstrates reversal of polarity displayed in the EEG recording shown in Figure 4.2 is best described as a:**
    A. Negativity at C4
    B. Negativity at P4
    C. Negativity at C4 and P4
    D. Positivity at C4 and negativity at P4

38. **The EEG recording shown in Figure 4.3 is normal.**
    A. True
    B. False

39. **Widespread slow activity in the right hemisphere is best seen with:**
    A. Longitudinal bipolar montage
    B. Transverse bipolar montage

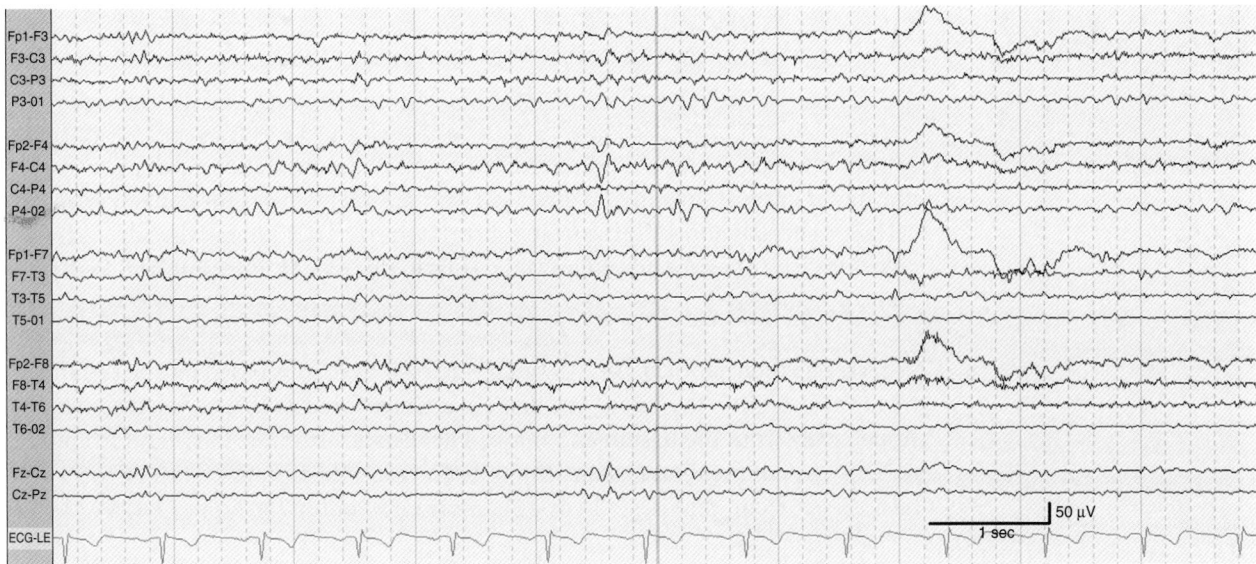

Figure 4.2.

C. Linked ear montage
D. Left ear reference montage

40. In a bipolar montage, an upward deflection is seen at the channel C4-P4. The potential at C4 is:
A. Positive
B. Negative
C. Neutral
D. Either positive or negative

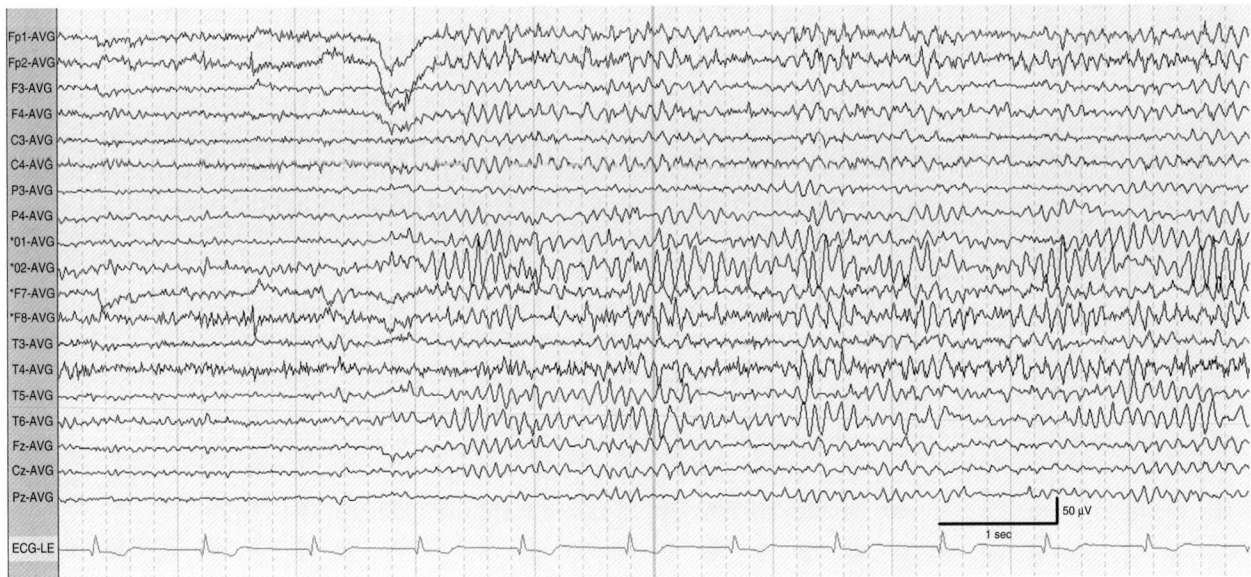

Figure 4.3.

ANSWERS

1. (D): Electrodes become polarized as a result of accumulation of charges that may favor current flow in one direction over another. This is more likely with certain materials and with recording direct current (DC) potentials. Silver-silver chloride electrodes are nonpolarizable. (**Fisch, pp. 23–24; Ebersole and Pedley 2003, pp. 50–54**)

2. (A): The international 10-20 system electrode system is based on the distance from the nasion to the inion. Fpz and Oz are measured at 10% of the distance respectively from the nasion and inion. The distance from Fz to Cz and from Cz to Pz is 20% of the total nasion–inion distance. So the distance from Fz to Pz will be 40% of 38 cm, that is, 15.2 cm. (**Ebersole and Pedley 2003, pp. 72–75**)

3. (C): The modified combinatorial nomenclature for the 10-10 electrode system uses the same landmarks as those of the 10-20 system for electrode placement. The 10-10 system expands the number of electrodes by adding new electrodes in between the 10-20 system electrodes. This is often used during long-term video-EEG monitoring where more accurate localization can be made. The number after the letter refers to the electrode position in relation to midline. In order to maintain this rule T3/T4 and T5/T6 electrodes had to be renamed. T7/T8 replace T3/T4 and P7/P8 replace T5/T6 (see Fig. 4.4). (**Fisch, pp. 551–553; Abou-Khalil and Misulis 2006, pp. 12–15**)

4. (D): Using a 30-Hz high-frequency filter (low pass filter) will attenuate any frequency >30 Hz. Higher frequencies are more attenuated. In this case, the highest frequency is 70 Hz and will be the most affected by that filter setting. (**Daube 2002, pp. 12–13**)

5. (D): The amplitude of a potential depends on its intensity, spatial orientation, and distance that separates it from the scalp. The resistance and capacitance of the structures between the generator of the potential and the scalp recording electrodes also affect inversely the amplitude. This is why scalp-recorded EEG is of lower amplitude than EEG activity recorded directly from the brain surface. (**Fisch, pp. 14–16; Daube 2002, pp. 28–39**)

6. (B): Scalp EEG signals record electrical activity originating from a generator through volume conduction. Near-field and far-field potentials respectively refer to potentials generated near and distant from the recording site. In the case of far-field potentials, the recording electrode captures a potential that has traveled through multiple heterogeneous media using volume conduction. As a result, the potential will have a wider scalp distribution and a lower voltage. (**Wyllie, Gupta and Lachhwani 2005, pp. 142–143; Daube 2002, pp. 184–185**)

7. (D): In analog or digital systems, increasing the paper speed or compressing the EEG recording will not have any impact on spike duration, amplitude, or frequency. Increasing the number of seconds displayed on the monitor may help better visualize focal slow activity, evolution of ictal discharges, and identification

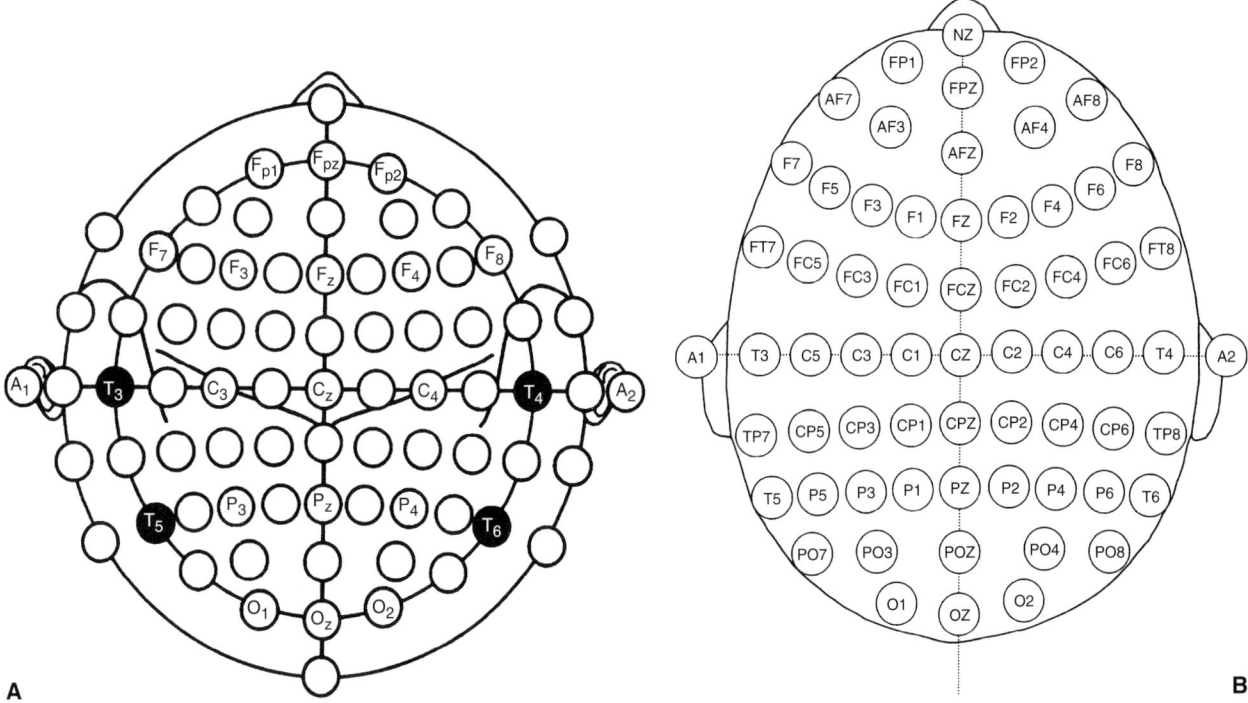

Figure 4.4. A: Electrode nomenclature of the 19 most commonly used electrodes, according to the International Federation of Clinical Neurophysiology 10-20 system. **B:** Electrode nomenclature in the International Federation of Clinical Neurophysiology 10-20 system with additional electrodes (this is the 10% system). (From Connolly MB, Sharbrough FW, Wong PKH. Electrical fields and recording techniques. In: Ebersole JS, Pedley TA, eds. *Current practice of clinical electroencephalography*, 3rd ed. Philadelphia: Lippincott Williams & Wilkins; 2003.)

of long-interval periodic discharges. (**Niedermeyer and Lopes Da Silva 1999, pp. 136–140**)

8. **(A):** The orientation of a cortical dipole in relation to the recording electrode is a major factor in determining the scalp electrical potential. This net potential is directly proportional to the cortical generator solid angle subtended by the recording electrode (see Fig. 4.5). (**Wyllie, Gupta and Lachhwani 2005, pp. 144–145; Niedermeyer and Lopes Da Silva 1999, pp. 146–148**)

9. **(B):** The paroxysmal depolarizing shift (PDS) represents the intracellular electrophysiological correlate of focal epileptiform discharges such as sharp waves and spikes. It consists of abnormal synchronous activation of multiple neurons at the cellular level causing a wave of depolarization. It is primarily due to the activation of high-frequency fast sodium channel potentials. (**Niedermeyer and Lopes Da Silva 1999, pp. 28–32; Daube 2002, pp. 57–59**)

10. **(D):** ECG artifact represents a far-field potential recorded at the scalp. The orientation of the cardiac field produces opposite polarity potentials at each ear. When the ears are linked, the opposite polarities of the ECG artifact will cancel out (especially if the ECG potentials are equal in voltage) leading to a less contaminated reference. ECG artifact is more pronounced in obese patients, children, and patients with short necks. (**Ebersole and Pedley 2003, pp. 275–276**)

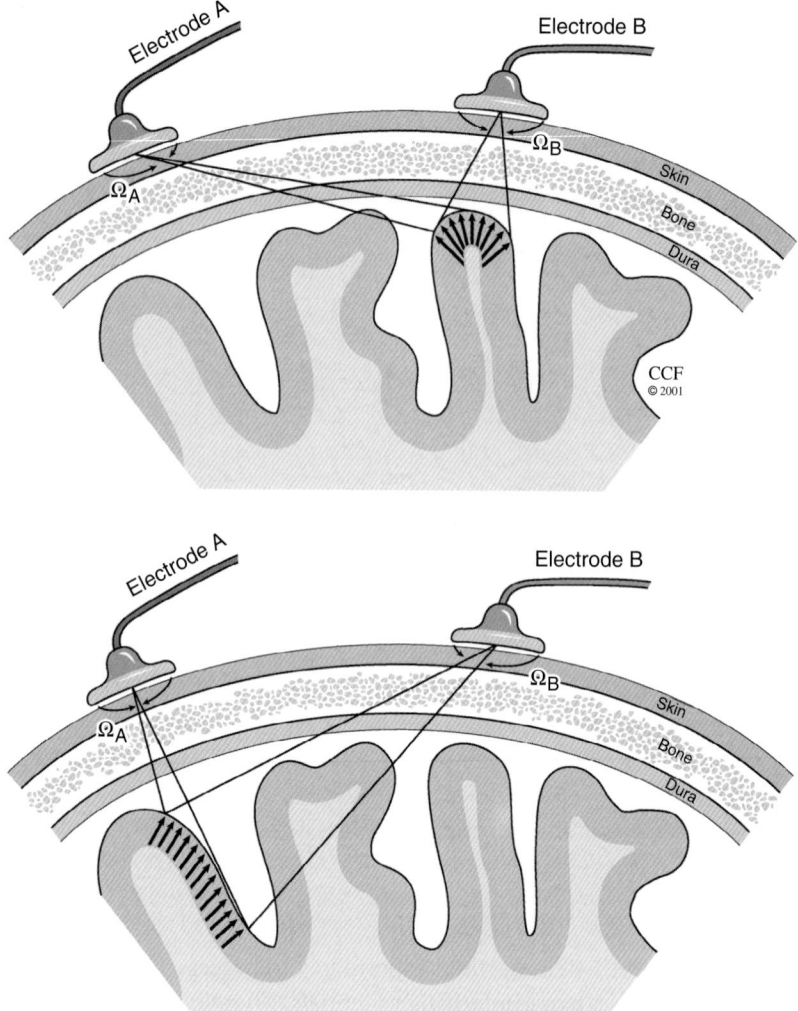

Figure 4.5. Use of the solid-angle rule to ascertain the signal measured on the scalp surface relative to the orientation of the dipole. *Top:* Surface electrode B sees a large electrical potential because of the orientation and proximity of the dipole layer, as borne out by the solid angle ΩB. *Bottom:* In this case, the potential seen by electrode A is actually lower than that measured by the more distant electrode B, because of the arrangement of the dipoles in the discharging region. The smaller solid angle, ΩA, is proportional to the voltage measured on the scalp. CCF, cerebral cortical function. (From Burgess RC, Iwasaki M, Nair D. Localization and field determination in electroencephalography and magnetoencephalography. In: Wyllie E, Gupta A, Lachhwani DK, eds. *The treatment of epilepsy: principles and practice*, 4th ed. Philadelphia: Lippincott Williams & Wilkins; 2006.)

11. (B): Bipolar montages both longitudinal (commonly referred to as *double banana*) or transverse link sequential pairs of electrodes into channels. Recorded potentials affecting one electrode will produce reversal of polarity in channels having this electrode in common. If two adjacent electrodes inside the chain are equally and maximally affected, the channel containing the two electrodes will show cancellation, but reversal of polarity will be noted across that channel. Reversal of polarity is not noted if the recorded potential (in this case a sharp wave) maximally involves the electrode at the beginning or end of the chain (end of chain phenomenon). **(Abou-Khalil and Misulis 2006, pp. 31–33)**

12. (B): All of the thalamic nuclei project directly to the cortex except the reticular nucleus of the thalamus. This nucleus receives input from the cortex and sends its output to other thalamic nuclei. It plays a major role in thalamocortical connectivity and is thought to be primarily responsible for the electrical synchronization that occurs with seizures. **(Fisch, pp. 10, 410–411)**

13. (B): Using the international 10-20 system, vertical eye movements are best seen in the frontopolar electrodes notably Fp1 and Fp2. Horizontal or lateral eye movements will be best recorded at F7 and F8 electrodes which are the nearest to the lateral side of the eyes. **(Fisch, pp. 107–110)**

14. (B): Several different techniques are used to attach electrodes to the scalp. In the case of long-term monitoring, the collodion technique offers a very stable recording with minimal artifact. In this technique, few drops of collodion are applied around the edge of the electrode. Stream of compressed air is used to dry the collodion. Adequate hyperventilation is needed for the safe use of collodion or acetone that removes collodion. **(Fisch, pp. 20–21)**

15. (A): Every EEG potential has a dipole with negative and positive ends. The EEG is best able to record radial dipoles that are perpendicular to the crown of the gyri. These dipoles are also called *vertical* or *perpendicular*. For these dipoles, only one end of the dipole is recorded on the scalp electrodes while the other end is directed deep into the brain. **(Fisch, pp. 87–90)**

16. (B): The voltage across the channel Fz-F3 is obtained by calculating the difference between the voltages at Fz and F3, i.e. $(-30\ \mu V) - (-70\ \mu V) = 40\ \mu V$. **(Abou-Khalil and Misulis 2006, pp. 30–38)**

17. (C): Activation of $GABA_A$ receptors open chloride channels causing hyperpolarizing potentials. Therefore, $GABA_A$ receptors are responsible for fast synaptic inhibition. Evidence suggests that impaired $GABA_A$ receptor function is the pathological basis of some inherited and acquired epilepsies. **(Wyllie, Gupta and Lachhwani 2005, p. 94)**

18. (C): Compared to the retina, the cornea is positively charged. This is illustrated when the EEG records eye closure. During eye closure, there is elevation of the eye (Bell's phenomenon) causing a positive potential (or downward deflection) in the frontopolar electrodes. The tip of the tongue is negatively charged as compared to its base. This is illustrated when the EEG records chewing. With mastication, there is rhythmic elevation of the tip of the tongue causing upward sinusoidal waves mimicking frontal rhythmic activity. **(Abou-Khalil and Misulis 2006, pp. 82–93)**

19. (B): Spike and sharp waves are transient epileptiform discharges defined by their duration. The duration of spikes is <70 ms whereas that of a sharp wave ranges from 70 to 200 ms. **(Abou-Khalil and Misulis 2006, pp. 48–49)**

20. (D): According to the international 10-20 system, the F7 and F8 electrodes lie over inferior frontal cortex, but most often record anterior temporal electrical activity. Depending on their field, they are referred to as *inferior frontal* or *anterior temporal electrodes*, and are sometimes referred to as the *frontotemporal electrodes*.

One way to determine the source of their activity is to evaluate involvement of adjacent electrodes. For example, if T3 is also involved, F7 is likely recording anterior temporal activity, and if F3 is also involved, F7 is more likely to be recording inferior frontal activity. (**Abou-Khalil and Misulis 2006, pp. 10–12**)

21. (D): A ground loop occurs when two or more instruments are attached to the patient and their ground levels are unequal. In that case, any leakage current from one instrument with higher voltage ground level may travel through the patient and then flow to the ground electrode of another instrument with a lower voltage creating ground a loop circle. This may produce a circuit of current flow referred to a *ground loop*, which increases the risk of patient electrocution. To avoid a ground loop circuit, one common ground is advisable especially in environments like the ICU, where multiple electrical equipments are attached to the patient. (**Fisch, p. 71**)

22. (A): Tetrodotoxin (TTX) is a nonprotein organic compound and one of the strongest paralytic toxins. It is a specific irreversible blocker of voltage-gated sodium (Na+) channels. It has been extremely valuable in Na+ channel identification and characterization in animal epilepsy models as well as in antiepileptic drug mechanisms. Similarly valuable was tetraethyl ammonia (TEA), a selective potassium channel blocker. (**Wyllie, Gupta and Lachhwani 2005, pp. 75, 78–79**)

23. (C): Baclofen is a $GABA_B$ agonist medication commonly used for spasticity. $GABA_B$ is a G-protein-coupled receptor that can open potassium or close calcium channels. Its receptors are presynaptic and postsynaptic. When activated, it modulates T-calcium channels and intrinsic bursting activities of the relay cell causing seizures. Benzodiazepines are $GABA_A$ agonists. (**Wyllie, Gupta and Lachhwani 2005, pp. 94–95**)

24. (D): Postsynaptic potentials and not action potentials contribute to the scalp-recorded electrical activity. This is because postsynaptic potentials are of long duration ranging from 10s to 100s of milliseconds. They involve a large area of the membrane surface and occur simultaneously in thousands of cortical cells arranged perpendicularly to the cortical surface. (**Daube 2002, p. 61**)

25. (D): The cortical surface involved is at least 6 cm^2 for a potential to be detected on scalp EEG. This is one reason why many simple partial seizures involving a small cortical area are often missed on scalp EEG. (**Fisch, pp. 13–16**)

26. (B): This EEG tracing is displayed in a referential average montage. After ictal onset (with the high voltage sharply contoured wave) there is rhythmic activity of opposite polarity at F7 and F8. Each positive deflection at F7 is preceded by a small spike representing lateral rectus muscle activity. The pattern is likely due to horizontal left-beating nystagmus. Lateral eye movements are best recorded at the electrodes closest to the eyes. The reversal of polarity is due to the movement of the positively charged cornea toward one electrode and away from the other electrode. Roving eye movements would be of much slower frequency. Eye flutter produces vertical eye movements best seen in the frontopolar electrodes. (**Ebersole and Pedley 2003, pp. 273–275**)

27. (**B**): Bancaud phenomenon is defined as a unilateral failure of attenuation of the occipital alpha rhythm to bilateral eye opening. It is an abnormal EEG finding. The side that does not attenuate is the abnormal side. Therefore this finding has a lateralizing value but not a localizing one. It is also nonspecific. (**Ebersole and Pedley 2003, p. 291**)

28. (**C**): Sphenoidal and zygomatic electrodes are added in patients with suspected temporal lobe seizures undergoing presurgical evaluation. Zygomatic electrodes are placed noninvasively as any other EEG lead over the skin. They are placed just below the zygomatic arch and record brain activity from the inferior-lateral temporal region. Sphenoidal electrodes are inserted to lie close to the foramen ovale. The electrode is attached to the tip of a spinal needle. The needle is introduced below the zygomatic arch, 2 cm anterior to the line between the tragus and the condyle of the mandible. The needle is pushed to a depth of 4 to 5 cm and then withdrawn leaving the wire in place. Sphenoidal electrodes record activity from the inferior-mesial temporal region (used for suspected mesial temporal lobe epilepsy). (**Abou-Khalil and Misulis 2006, pp. 12–13**)

29. (**A**): Glossokinetic artifact refers to chewing during the normal act of eating. The tip of the tongue is negatively charged compared to its base. During chewing, tongue movement produces rhythmic low frequency sinusoidal waveforms superimposed on muscle activity. When filtered, this glossokinetic artifact will mimic projected rhythms such as diffuse, bilateral frontal or temporal rhythmic delta activity. In nonresponsive patients, glossokinetic artifact may be distinguished from pathological projected rhythm by the presence of associated electromyogram (EMG) artifact from the temporalis and frontalis muscles, and its disappearance during sleep. In addition glossokinetic potentials will have a higher amplitude in the infraorbital electrodes in comparison with frontopolar electrodes. (**Blume, Kaibara and Young 2002, pp. 27–29; Abou-Khalil and Misulis 2006, pp. 92–93**)

30. (**C**): Accurate description of wave repetition is essential in EEG analysis. Rhythmic activity recurs at a rate that is equal to the waveform frequency. An example of rhythmic activity is frontal intermittent rhythmic delta activity (FIRDA) or temporal intermittent rhythmic delta activity (TIRDA). Periodic activity recurs at a lower rate than the waveform frequency because the recurring waveforms are separated by intervening activity. Example of periodic activity is periodic lateralized epileptiform discharges (PLEDs). (**Abou-Khalil and Misulis 2006, pp. 39–43**)

31. (**B**): EEG frequency reduction (slow activity) is a function of white matter disturbance whereas voltage or amplitude reduction (attenuation) is a function of cortical disturbance. The mechanism of focal slowing is thought to be due to partial cortical deafferentation from subcortical structures rather than due to metabolic changes. This could be due to functional or anatomic deafferentiation

of the cortex as may be seen in toxic-metabolic encephalopathies or structural brain lesions. Thalamic desynchronization rather than synchronization may also play a role in focal slowing such as polymorphic delta activity. (**Fisch, pp. 354–358; Ebersole and Pedley 2003, pp. 313–316**)

32. (C): Several technical standards are commonly used in recording routine EEG studies. These are:
Electrode placement according to the 10-20 international system
Impedance of every electrode not exceeding 5,000 Ω
Recording simultaneously from minimum of eight electrodes
Alternating a minimum of three montages: longitudinal and transverse bipolar and one reference montage
Sensitivity of 5 to 10 μV per mm
Low frequency filter of 1 Hz and high-frequency filter of 70 Hz
Calibration at beginning and end of recording
Speed of 30 mm per second
A minimum of 20 minutes of artifact-free recording in the awake state with frequent brief periods of eye opening and closure (**Fisch, pp. 94–99**)

33. (B): In order to better appreciate focal slow activity, the reader can perform three changes to the settings of the EEG recording. Compressing the EEG by decreasing the speed of recording and increasing the sensitivity will permit better appreciation of frequency differences. Also increasing the time constant (inversely proportional to the low frequency filter) will allow more slow activity to be displayed. (**Fisch, pp. 46–54; Abou-Khalil and Misulis 2006, pp. 21–22**)

34. (C): As a general rule, the reference that is the least contaminated by the studied field will provide the best spatiotemporal representation. If the activity of interest is confined to the temporal lobe, ear references are likely contaminated by the temporal lobe activity causing reduction or even cancellation of the activity in the temporal leads. The average reference (average of all electrodes except Fp1 and Fp2) is usually suitable for recording temporal lobe activity, but may also be contaminated by the activity of interest if it has a high amplitude and a wide field. The vertex reference is the furthest from the temporal lobe and will provide the least contaminated reference. In sleep the vertex reference is not suitable because of contamination with high-amplitude activity of physiological sleep potentials (vertex waves, K-complexes). (**Fisch, pp. 78–84; Abou-Khalil and Misulis 2006, p. 189**)

35. (B): Like hyperventilation, photic stimulation is another activation technique that is performed in routine EEG recordings. It produces occipital visual evoked potentials produced by flashlights. It is not seen in neonates. It is largest at frequencies close to the patient's occipital alpha frequency (8 to 15 Hz) and higher amplitude during eye closure. (**Fisch, pp. 223–225**)

36. (D): Sharp waves and spikes are abnormal sharp transients strongly associated with epilepsy. They are referred to as *epileptiform discharges*. Several features can help distinguish epileptiform from nonepileptiform sharp transients, including polyphasic morphology, high amplitude, after-going slow waves. Other helpful features include asymmetrical morphology, disruption of the background, and emanation from an abnormal background. These features are not absolute. **(Abou-Khalil and Misulis 2006, pp. 48–50)**

37. (C): This EEG tracing (Fig. 4.2) is displayed in a longitudinal bipolar montage. There is reversal of polarity across the channel C4-P4. According to the rules of polarity, the potential is negative at both C4 and P4 electrodes and its cancellation across that channel means that the two electrodes are equally affected. In summary, there is a negative reversal of polarity across a zone of equipotentiality between C4 and P4. (See Fig. 4.6). **(Abou-Khalil and Misulis 2006, pp. 17–18, 30–38)**

38. (A): This EEG tracing (Fig. 4.3) is displayed in an average referential montage. Following eye closure, there is an asymmetrical alpha rhythm, but usually with <50% asymmetry. It is common to see a higher-amplitude posterior rhythm on the right side because of differences in skull thickness. When the difference exceeds 50%, the lower amplitude side is considered abnormal. (See Table 4.1). **(Fisch, pp. 187–188; Blume, Kaibara and Young 2002, pp. 361–362)**

39. (D): The principle of using a reference that is not contaminated by the abnormal activity makes the left ear the reference of choice. Linked ear montages would include the ear ipsilateral to the slowing. Bipolar montages may mask the widespread right hemisphere slowing when electrodes are equally affected. **(Wyllie, Gupta and Lachhwani 2005, pp. 150–155)**

40. (D): In a bipolar montage, the active electrode can be either the first or second input. Applying the rules of polarity to that concept, the potential at C4 can be either positive or negative (see Fig. 4.7, Table 4.2). **(Abou-Khalil and Misulis 2006, pp. 17–18; Ebersole and Pedley 2003, pp. 75–76)**

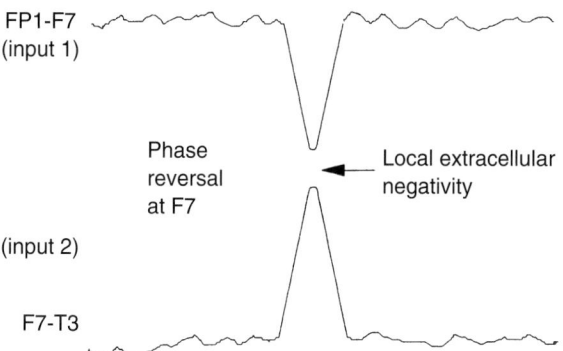

Figure 4.6. Schematic drawing of a phase reversal, representing site of local extracellular negativity, which in this illustration is at the F7 electrode. (From Litt B, Cranstoun SD. Engineering principles. In: Ebersole JS, Pedley TA, eds. *Current practice of clinical electroencephalography*, 3rd ed. Philadelphia: Lippincott Williams & Wilkins; 2003.)

TABLE 4.1 Patterns of Electroencephalogram (EEG) Rhythms Divided by Frequency

Rhythm	Frequency (Hz)	Normal Patterns	Abnormal Patterns
Delta	<4	Slow wave sleep (diffuse)	Irregular delta activity (focal or diffuse)
		Hyperventilation (diffuse with frontal predominance)	Rhythmic delta activity (temporal—TIRDA, frontal—FIRDA, and occipital—OIRDA)
		Posterior slowing of the youth (occipital)	Delta frequency seizures
Theta	4–8<	Drowsiness (diffuse)	Irregular theta activity (focal or diffuse)
		Slow alpha variant (occipital)	Theta frequency seizures
		Rhythmic midtemporal theta of drowsiness (RMTD) (temporal)	
		Subclinical rhythmic EEG discharge of adults (SREDA) (parietal)	
Alpha	8–13	Posterior rhythm (occipital)	Alpha coma (anterior predominance)
		Mu rhythm (central)	Alpha frequency seizures
		Third rhythm (temporal)	
Beta	>13	Drowsiness in children (diffuse)	Breach rhythm (over skull defect)
		Fast alpha variant (occipital)	Medication induced (diffuse with frontal predominance)
			Beta frequency seizures

TIRDA, temporal intermittent rhythmic delta activity; FIRDA, frontal intermittent rhythmic delta activity; OIRDA, occipital intermittent rhythmic delta activity.

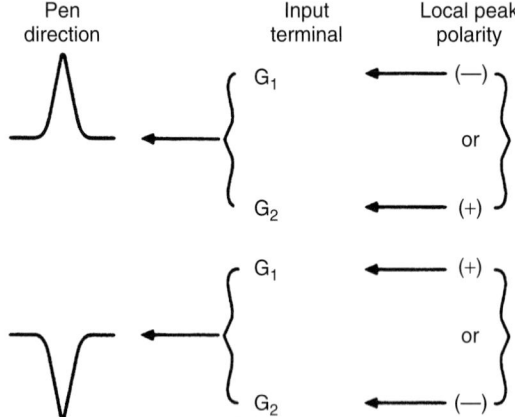

Figure 4.7. According to the standard polarity convention, an upward signal deflection results if input 1 is more negative than input 2 or if input 2 is more positive than input 1. Conversely, a downward signal deflection results if input 1 is more positive than input 2 or if input 2 is more negative than input 1. (From Connolly MB, Sharbrough FW, Wong PKH. Electrical fields and recording techniques. In: Ebersole JS, Pedley TA, eds. *Current practice of clinical electroencephalography*, 3rd ed. Philadelphia: Lippincott Williams & Wilkins; 2003.)

TABLE 4.2 Rules of Polarity

Input 1	Input 2	EEG activity
Up	Down	Negative
Down	Up	Positive

References

1. Abou-Khalil B, Misulis Karl. *Atlas of EEG and seizure semiology*. Elsevier Science; 2006.
2. Blume WT, Kaibara M, Young GB. *Atlas of adult electroencephalography*, 2nd ed. Lippincott Williams & Wilkins; 2002.
3. Daube J. *Clinical neurophysiology*, 2nd ed, Oxford Press; 2002.
4. Ebersole J, Pedley T. *Current practice of clinical electroencephalograph*, 3rd ed. Lippincott Williams & Wilkins; 2003.
5. Fisch B. *Fisch and Spehlmann's EEG printer: principles of digital and analog EEG*, 3rd ed. Elsevier Science; 1999.
6. Niedermeyer E, Lopes Da Silva F. *Electroencephalography: basic principles, clinical applications, and related fields*, 4th ed. Lippincott Williams & Wilkins; 1999.
7. Wyllie E, Gupta A, Lachhwani DK. *The treatment of epilepsy: principles and practice*, 4th ed. Lippincott Williams & Wilkins; 2005.

CHAPTER 5

Neonatal and Pediatric Electroencephalography

QUESTIONS

1. **All of the following is true about neonatal electroencephalographic (EEG) electrode placement except:**
   A. Typically, nine scalp electrodes are used
   B. F1 or Fp3 and F2 or Fp4 are the most frontal electrodes placed in accordance to the 10-20 international system
   C. The longitudinal bipolar montage is the most commonly used montage for recording
   D. Electrooculogram, electromyogram, and respiratory monitoring are often added

2. **In full-term infants, sleep spindles become apparent at the age of:**
   A. Birth
   B. 1 to 3 months
   C. 6 to 12 months
   D. 12 to 24 months

3. **Canavan disease will likely produce:**
   A. Generalized spike-and-wave discharge
   B. Diffuse polymorphic delta activity
   C. Periodic lateralized epileptiform discharges (PLEDs)
   D. Temporal intermittent rhythmic delta activity (TIRDA)

4. **The EEG of patients with Rasmussen syndrome is likely to reveal:**
   A. Absence status
   B. Focal electrographic status
   C. Generalized attenuation
   D. Burst suppression pattern

5. **The EEG of a 6-year-old boy with progressive regression of speech will more likely show abnormalities during:**
   A. Wakefulness
   B. Drowsiness
   C. Non–rapid eye movement (NREM) sleep
   D. REM sleep

6. **The EEG tracing shown in Figure 5.1 is that of a healthy 6-year-old girl. Which of the following medical conditions may exacerbate this EEG pattern?**
   A. Acute intermittent porphyria
   B. Alzheimer's dementia
   C. Reye syndrome
   D. Hashimoto's thyroiditis

7. **Hypnagogic hypersynchrony is most commonly seen in:**
   A. Neonates
   B. Children
   C. Menstruating women
   D. Elderly with dementia

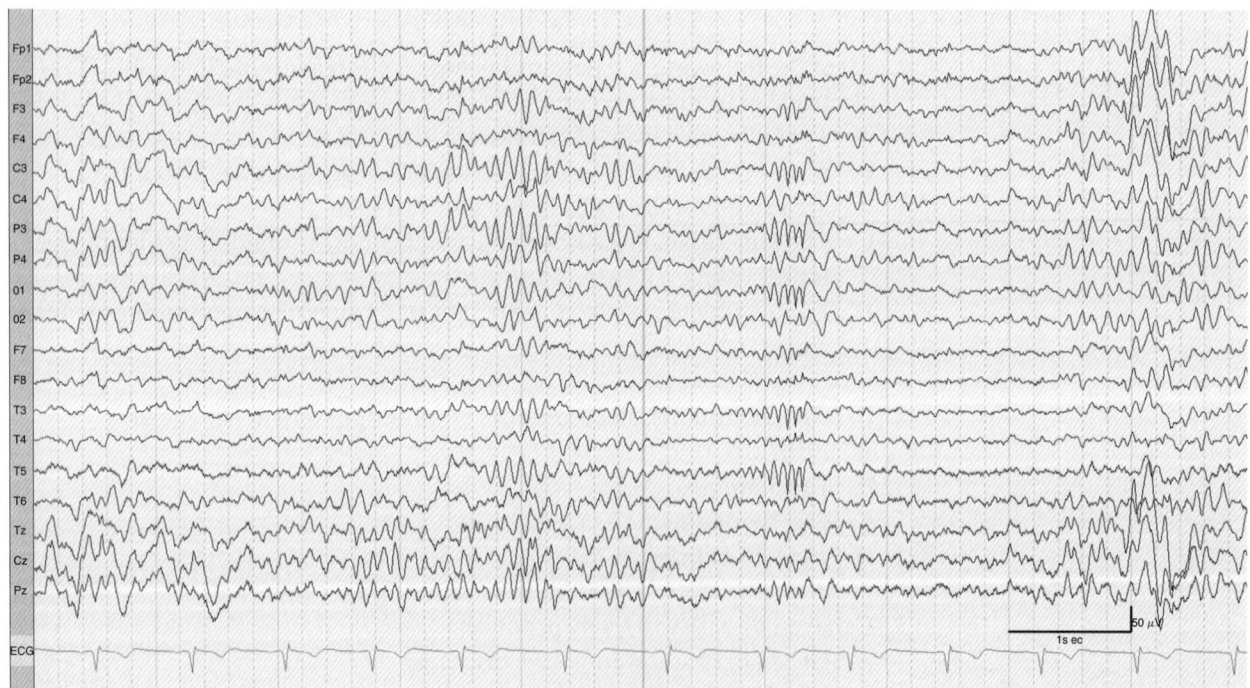

Figure 5.1.

8. The EEG tracing shown in Figure 5.2 belongs to a normal child. On the basis of this EEG, the age of this child is most likely:
   A. Neonate
   B. 3 months to 2 years old
   C. 2 to 5 years old
   D. 6 to 8 years old

9. West syndrome is characterized by:
   A. Myoclonic seizures
   B. Infantile spasms
   C. Absence seizures
   D. All of the above

10. The recommended length of a neonatal EEG is:
    A. 20 minutes
    B. 30 minutes
    C. 60 minutes
    D. 90 minutes

11. At which gestational age is wakefulness distinguished from sleep?
    A. 28 to 30 weeks
    B. 30 to 34 weeks
    C. 34 to 36 weeks
    D. 36 to 38 weeks

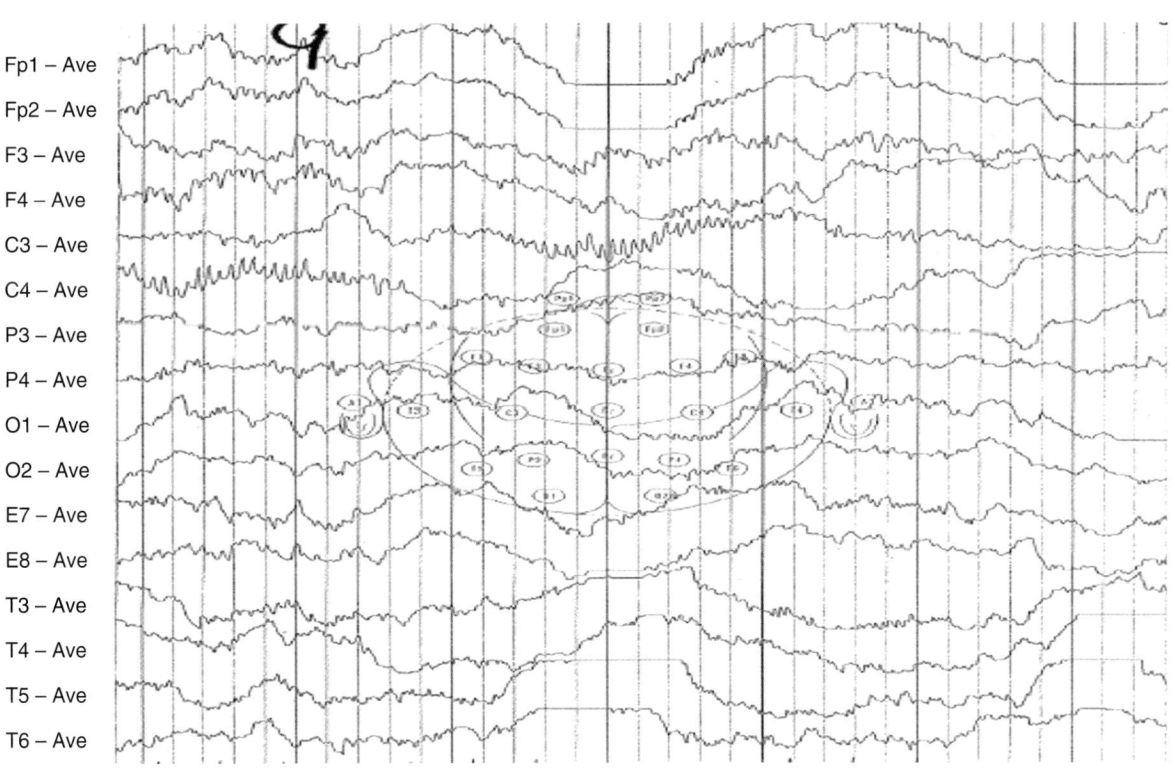

Figure 5.2.

12. The EEG recording of a 15-year-old male patient with progressive cognitive decline reveals regular periodic complexes every 8 seconds. The likely diagnosis is:
    A. Lithium toxicity
    B. Rasmussen syndrome
    C. Subacute sclerosing panencephalitis
    D. Creutzfeld-Jacob disease

13. A 7-year-old male has nocturnal seizures consisting of strange guttural noise and left face contraction followed by generalized jerking. The highest yield EEG is during:
    A. Hyperventilation
    B. Photic stimulation
    C. Wakefulness
    D. Sleep

14. The EEG recording shown in Figure 5.3 is most consistent with the diagnosis of:
    A. West syndrome
    B. Lennox-Gastaut syndrome
    C. Benign epilepsy with centrotemporal spikes
    D. Juvenile myoclonic epilepsy

15. The EEG of a 6-year-old boy with congenital blindness will most likely show:
    A. Slowing of the occipital rhythm
    B. Occipital spikes
    C. A and B
    D. Congenital blindness does not affect the EEG

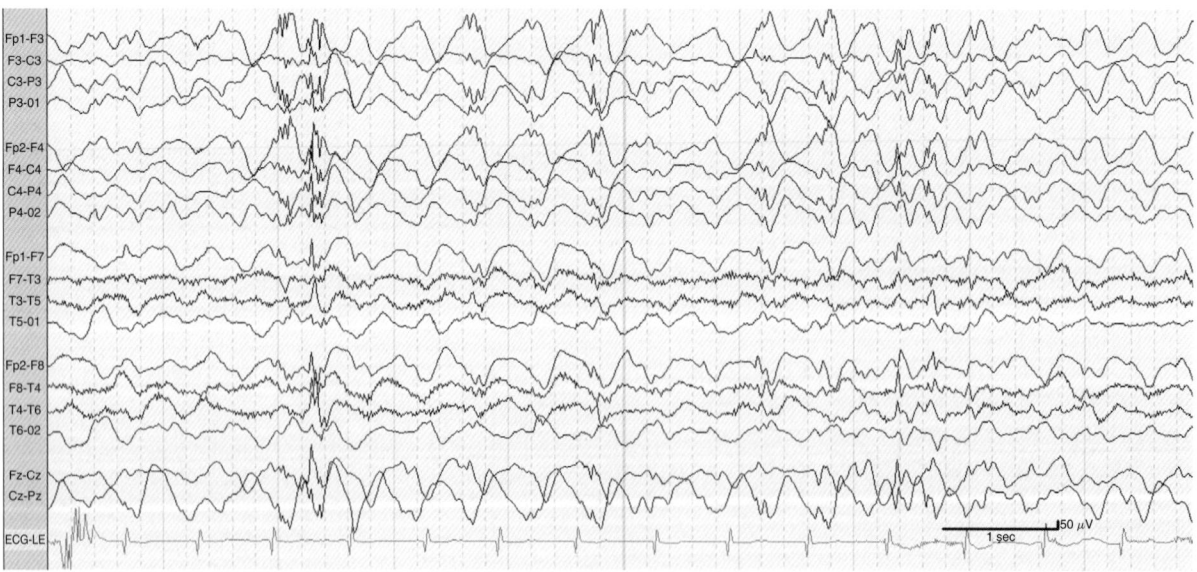

Figure 5.3.

16. The EEG recording shown in Figure 5.4 is most consistent with the diagnosis of:
    A. Multifocal epilepsy
    B. Benign rolandic epilepsy with centrotemporal spikes
    C. Symptomatic generalized epilepsy
    D. Idiopathic generalized epilepsy

17. A 12-year-old male with severe autism has bursts of high-amplitude 3 Hz sharply contoured delta activity during hyperventilation. This is most consistent with the diagnosis of:
    A. Symptomatic generalized epilepsy
    B. Lennox-Gastaut syndrome
    C. Encephalopathy
    D. Normal finding

18. Which of the following EEG properties of benign occipital epilepsy of childhood is true?
    A. Interictal discharges are always seen in the occipital regions
    B. Extraoccipital epileptiform discharges are not present
    C. Ictal discharges consist of low-voltage fast activity in late onset benign occipital lobe epilepsy
    D. Discharges are attenuated with eye opening

19. Benign EEG patterns in a 7-year-old child include all of the following except:
    A. Frontal intermittent rhythmic delta activity (FIRDA)
    B. 14 and 6 Hz positive spikes
    C. Rare frontal sharp waves
    D. Delta activity intermingled in the occipital rhythm

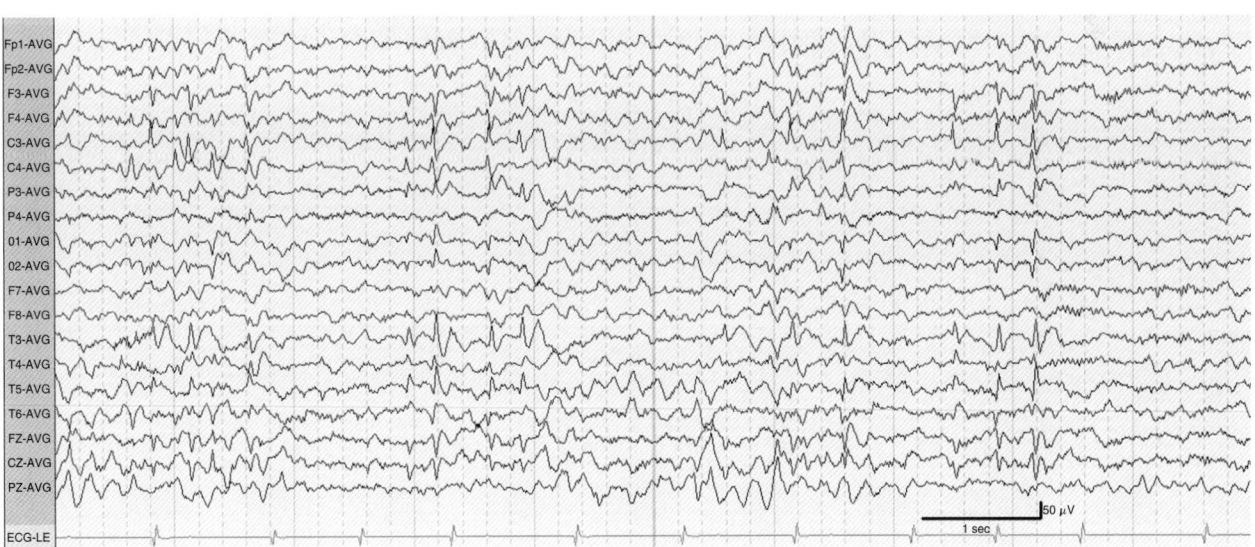

Figure 5.4.

20. In a 28-week preterm infant, the EEG shows a pattern of:
    A. Tracé alternant
    B. Tracé discontinu
    C. Tracé continu
    D. Tracé parallel

21. A 12-year-old boy has EEG attenuation in the left hemisphere. Among the following, which condition is most likely to show this finding?
    A. Tuberous sclerosis
    B. Neurofibromatosis I
    C. Sturge-Weber
    D. Von-Hippel Lindau

22. Adult alpha range frequency of the posterior rhythm is attained at the age of:
    A. 2 to 3 years
    B. 4 to 6 years
    C. 8 to 12 years
    D. 14 to 16 years

23. All of the following is true regarding neonatal seizures except:
    A. Most ictal discharges originate in the central regions
    B. Ictal discharges are often generalized due to immaturity of the brain
    C. Ictal discharges consist of rhythmic activity of various morphologies lasting for 10 seconds or more
    D. Electrical seizures are rare before the age of 34 to 35 weeks of conceptual age

24. The EEG recording shown in Figure 5.5 is that of a 14-year-old male with spells of "spacing out". This EEG is most consistent with the diagnosis of:
    A. Partial epilepsy with a right frontal focus
    B. Partial epilepsy with bilateral independent frontal foci

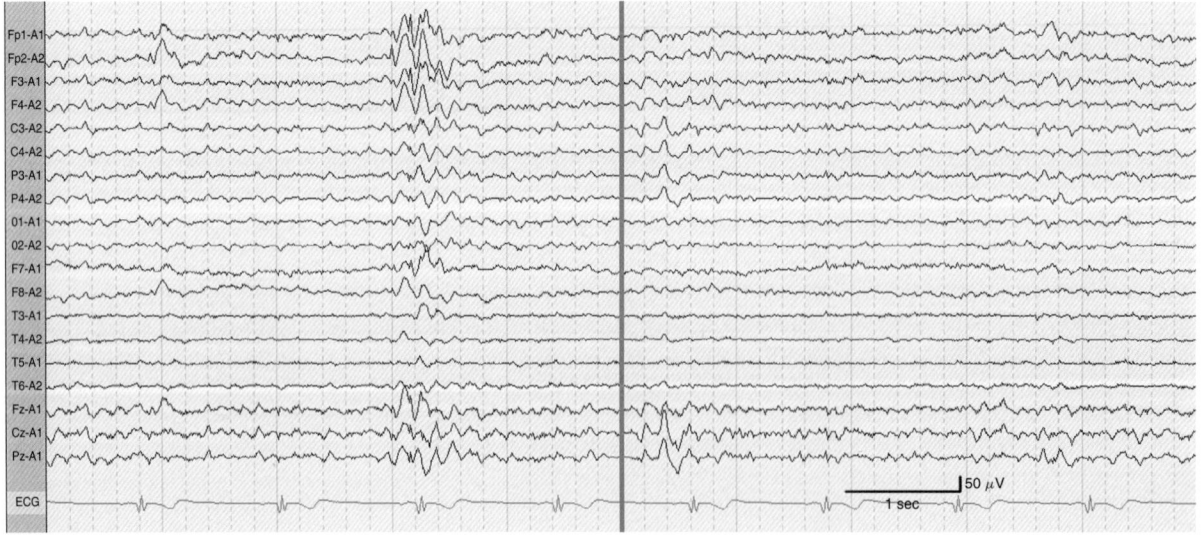

Figure 5.5.

C. Idiopathic generalized epilepsy
D. Normal EEG variant during drowsiness

25. **In comparison to a full-term infant, the EEG of a premature infant will have all of the following properties except:**
    A. More frequent sharp transients
    B. More asynchrony
    C. More quiet sleep
    D. More active sleep

26. **Which of the following is not true about delta brushes?**
    A. They are mainly seen between the ages of 32 and 36 weeks of gestation
    B. They can persist up to the age of 42 to 44 weeks of conceptual age
    C. They are predominantly seen in the frontal regions
    D. They are more frequent during sleep

27. **In a 36-week-old infant, frontal sharp transients indicate:**
    A. Frontal epileptogenicity
    B. Encephalopathy
    C. Focal frontal dysfunction
    D. None of the above

28. **Distinct vertex waves are formed by age:**
    A. Birth
    B. 1 to 3 months
    C. 3 to 5 months
    D. 5 to 9 months

29. **In comparison with older children, younger children may have more hyperventilation-induced slow activity over:**
    A. The anterior head region
    B. The posterior head region
    C. The central head region
    D. Less slow activity over all

30. **In the first 3 months of life, the predominant EEG rhythm during wakefulness is:**
    A. Central rhythm of 5 to 6 Hz
    B. Occipital rhythm of 3 to 4 Hz
    C. Central rhythm of 6 to 8 Hz
    D. Occipital rhythm of 7 to 8 Hz

31. **Quiet sleep of a normal full-term neonate consists of all of the following except:**
    A. Tracé alternant
    B. Tracé discontinu
    C. High-voltage slow pattern
    D. Delta brushes

32. The EEG recording shown in Figure 5.6 is that of a healthy full-term baby during sleep. The sleep stage being recorded is most likely during:
    A. Active sleep
    B. Quiet sleep
    C. Rapid eye movement (REM) sleep
    D. Drowsiness

33. Infantile spasms are electrographically associated with:
    A. Burst suppression pattern
    B. Electrodecremental pattern
    C. Electrocerebral silence
    D. Repetitive sharp waves

34. Which of the following is true about seizures with electrodecremental response?
    A. Usually arises from a normal background
    B. Often associated with tonic seizures
    C. It is more likely in idiopathic generalized epilepsy
    D. It ends as soon as the clinical seizure ends

35. A premature infant with apneic spells has an EEG showing frequent multifocal sharp waves. This finding is most consistent with the diagnostic of:
    A. Multifocal potential epileptogenicity
    B. Nonspecific generalized cerebral dysfunction
    C. Hypsarrythmia
    D. Normal finding

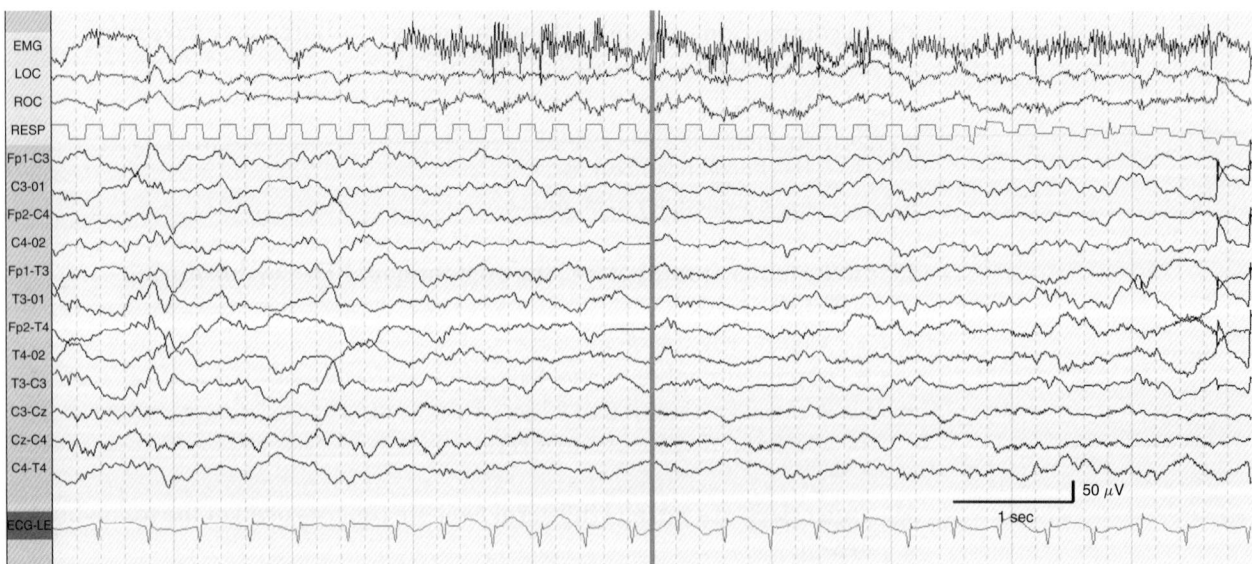

Figure 5.6.

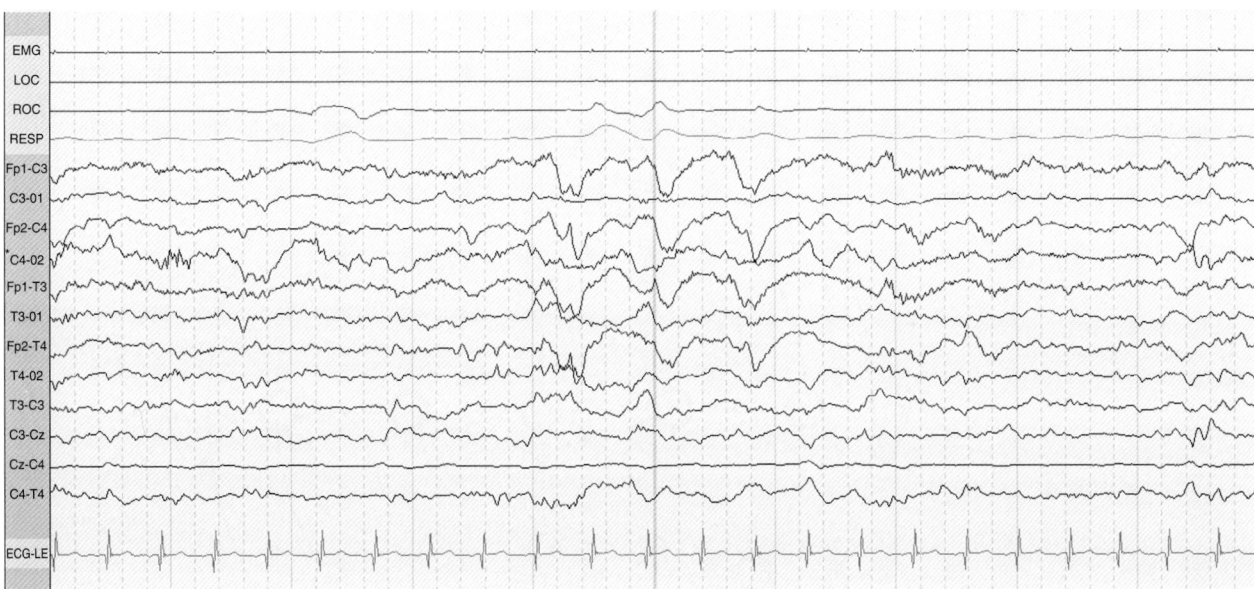

Figure 5.7.

36. The EEG shown in Figure 5.7 is that of a 37-week-old baby boy during sleep. The EEG findings are most consistent with:
   A. Normal EEG
   B. Encephalopathy
   C. Bifrontal independent partial epilepsy
   D. Brain dysmaturity

ANSWERS

1. **(B):** Neonatal EEG electrode placement is mildly modified from the 10-20 system to accommodate for the smaller head circumference and the relatively immature frontal lobes. Every EEG laboratory uses its own electrode system, but typically nine electrodes are used (F1, F2, C3, C4, Cz, T3, T4, O1, and O2). F1/Fp3 and F2/Fp4 are the frontal electrodes measuring 20% from the nasion. Also, typically a longitudinal bipolar montage is used for the whole recording in order to identify state changes and characterize reactivity and abnormalities. Commonly, infraorbital and submental electrodes are used to help identify state changes in addition to respiration monitoring which can help distinguish respiration artifact from cerebral activity (Fig. 5.8). (**Mizrahi, Hrachovy and Kellaway 2003, pp. 5–10; Ebersole and Pedley 2003, pp. 161–162**)

2. **(B):** Sleep spindles are formed at the age of 1 to 3 months in term infants, maximally in the central regions. They are initially bilateral but asynchronous until the age of 2 years when they become synchronous. In the first year, their morphology may be arciform or comb-like maturing in the later stages of life.

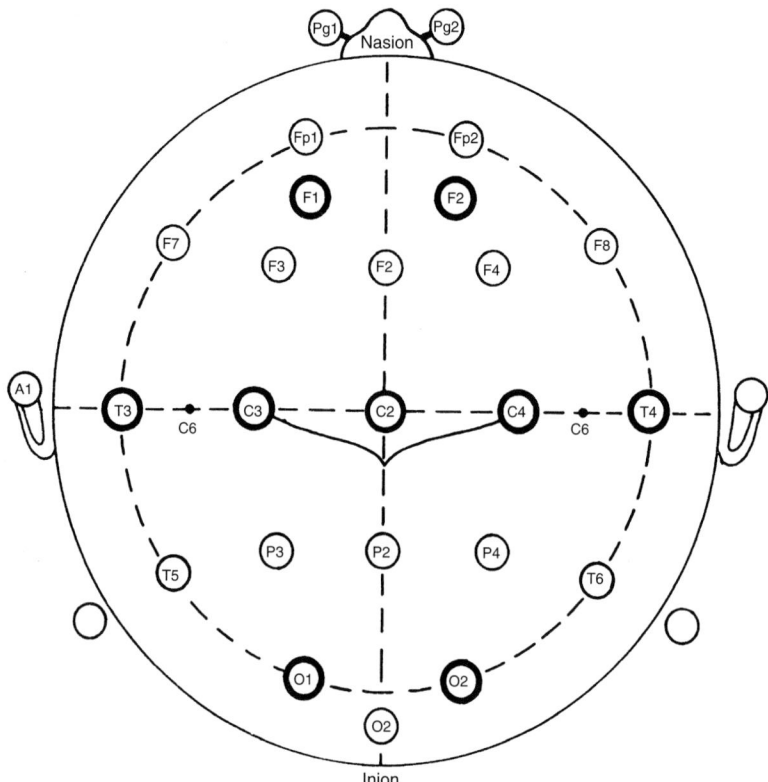

Figure 5.8. Electrode placement for the neonatal electroencephalogram designated by *bolded circles* (From Mizrahi EM, Hrachovy RA, Kellaway P. *Atlas of neonatal encephalography*. Philadelphia: Lippincott Williams & Wilkins, 2004.)

They are also more frequent in the early years of life during NREM sleep. Spindle duration is approximately 2.5 seconds at 3 months and 0.75 seconds at 22 months. Spindle frequency remains constant at 12 to 16 after age 5. (**Blume and Kaibara 1999, pp. 5–6, 124–128; Daube 2002, p. 114**)

3. (**B**): Canavan disease like all leukodystrophies affects the white matter. In the case of Canavan disease, there is diffuse white matter disease, which will produce diffuse EEG slow activity such as continuous polymorphic delta activity. Polymorphic delta activity consists of high-amplitude delta waves with irregular shapes and variable duration and frequency. This EEG pattern represents a nonspecific diffuse cerebral dysfunction and is often nonepileptiform in nature. For example, it can be seen in toxic, metabolic states as well as brainstem lesions such as the diencephalon.

PLEDs and TIRDA may also represent nonspecific EEG patterns but they are often associated with focal lesions. Generalized spike-and-wave discharges are seen in generalized epilepsies. (**Ebersole and Pedley 2003, pp. 294–295, 515**)

4. (**B**): Rasmussen syndrome is characterized by progressive intellectual deterioration and recurrent seizures of different types that usually progress into the form of epilepsia partialis continua (in 50% of the cases), also known as *focal motor status epilepticus*. Clinically, the seizures are simple motor with possible progression into complex partial or secondarily generalized seizures. Because this disease affects the cortical and subcortical areas of one hemisphere, the interictal

EEG may show lateralized slow activity and abundant epileptiform discharges. Hemispherectomy is usually the treatment of choice. **(Wyllie, Gupta and Lachhwani 2005, pp. 272–273)**

5. **(C):** The clinical history may be consistent with the diagnosis of acquired epileptic aphasia commonly referred to as *Landau-Kleffner syndrome*. This condition affects young children (3 to 9 years of age) causing regression of speech, personality changes, with or without seizures.

   The EEG often shows focal, bilateral independent, or generalized epileptiform activity in the form of spikes or spike-and-wave discharges. These discharges are markedly activated during NREM sleep when they become extremely frequent. These discharges can be missed in wakefulness, drowsiness, or REM sleep recordings. **(Ebersole and Pedley 2003, pp. 554–556)**

6. **(C):** This EEG tracing (Fig. 5.1) is displayed in an average referential montage. The presence of vertex waves and low-voltage mixed theta and delta activity suggest a drowsy or light sleep state. There is a burst of sharply contoured waveform of 14 and 6 Hz frequency. This EEG pattern is referred to as *14 and 6 positive spikes* and is most prominently seen in the posterior temporal region of healthy children. This pattern is most prominent in drowsiness and light sleep.

   Interestingly, this EEG pattern is excessively seen in children with Reye syndrome (acute childhood toxic encephalopathy). Reye syndrome is always associated with hepatic dysfunction but unlike in adults it is not strongly associated with triphasic waves. **(Ebersole and Pedley 2003, pp. 359–361; Blume, Kaibara and Young 2002, pp. 168–169)**

7. **(B):** Hypnagogic hypersynchrony refers to a normal EEG pattern seen at the beginning of drowsiness. It consists of abrupt onset of rhythmical and bisynchronous 4 to 5 Hz high-amplitude waves, often with a wide distribution. It is first seen in children aged 3 months and gradually vanishes by age 12 years. It can also occur at the end of drowsiness when it is referred to as *hypnopompic hypersynchrony*. **(Fisch 1999, pp. 172–173; Ebersole and Pedley 2003, pp. 132–140)**

8. **(B):** This EEG tracing (Fig. 5.2) is displayed in an average referential montage. There are symmetrical but asynchronous sleep spindles, maximally seen in the central electrodes. Sleep spindles are initially formed bilaterally but asynchronously by 1 to 3 months of age. They are predominantly synchronous by the age of 2 years. If they remain asynchronous, they are suggestive of structural abnormalities such as agenesis of the corpus callosum. **(Ebersole and Pedley 2003, pp. 144–145; Blume and Kaibara 2002, pp. 123–126)**

9. **(B):** West syndrome consists of a triad of infantile spasms, developmental delay, and a grossly abnormal EEG pattern termed *hypsarrythmia*. It usually affects infants between the ages of 4 to 8 months. In most cases, West syndrome is associated with serious neurologic abnormalities, including genetic, structural, metabolic, or infectious. **(Ebersole and Pedley 20003, pp. 540–544)**

10. **(C):** The recommended length of a neonatal EEG recording is 60 minutes. This is because the neonatal sleep cycle is approximately 45 to 60 minutes in a

full-term neonate. This will allow the sampling of all the neonatal sleep stages that approximately include 25 minutes of active sleep, 20 minutes of quiet sleep, and 15 minutes of intermediate sleep. Unlike neonates, adults have a typical sleep cycle of 80 to 120 minutes. (**Fisch 1999, pp. 159, 205–206**)

11. (B): Wakefulness or active sleep can be distinguished from quiet sleep by the gestational age of 30 to 34 weeks. Sleep is divided into active and quiet sleep. Wakefulness and active sleep are similar to REM sleep in which the muscle tone is low, respiration and ECG are irregular, and grimaces and twitches are common. At the age of 30 to 34 weeks, the active sleep EEG consists of a relatively more continuous pattern with shortening of the interburst interval duration.

On the other hand, quiet sleep which is similar to non-REM sleep consists of reduced eye and muscle movements with regular ECG and respiration rate. Quiet sleep EEG remains discontinuous with prolonged interburst interval duration known as *tracé discontinu*. (**Mizrahi, Hrachovy and Kellaway 2003, pp. 56–61; Eebersole and Pedley 2003, pp. 183–184**)

12. (C): Subacute sclerosing panencephalitis (SSPE) is a slowly progressive encephalitis characterized by cognitive decline, behavioral changes, myoclonus, seizures, movement disorder, and visual loss. SSPE is caused by central nervous system (CNS) involvement with the measles virus. It carries a very poor prognosis.

In this condition, the periodic EEG complexes are pathognomonic in the appropriate clinical setting. Complexes are of high voltage; usually consisting of diffuse or symmetrical sharp waves or sharply contoured waves that are persistent throughout the recording. The interval varies from 4 to 20 seconds, being longer in the early stages of the disease and shorter with disease progression. Compressing the EEG by decreasing the speed of recording may help in appreciating the regularity of the long interval complexes. When present, myoclonic jerks tend to occur in deep sleep and almost always coincide with the complexes. (**Blume and Kaibara 1999, pp. 283, 368–370**)

13. (D): The age and description of the seizures is consistent with benign epilepsy with centrotemporal spikes (BECTS) also known as *benign rolandic epilepsy*. Onset of seizures occurs between the ages of 4 to 10 years. Clinically, seizures are common during sleep (80% of times) consisting of unilateral tonic or clonic activity of the face and excessive salivation often followed by secondarily generalization.

Interictal EEG demonstrates characteristic central midtemporal spikes. Sleep dramatically activates these discharges. Discharges are more common in slow wave sleep than REM sleep. About one third of patients have interictal discharges only during sleep. (**Ebersole and Pedley 2003, pp. 526–532**)

14. (B): This EEG tracing (Fig. 5.3) is displayed in a longitudinal bipolar montage. The background activity is disorganized with frequent 1.5 to 2 Hz generalized spike-and-wave discharges. These, referred to as *slow spike-and-wave discharges*, are seen in Lennox-Gastaut syndrome (LGS), a form of symptomatic generalized epilepsy. Slow spike-and-wave discharges can be symmetrical, can have shifting predominance, or even be consistently asymmetrical. They are commonly seen in the anterior and central regions and are typically seen during NREM sleep. Patients with LGS usually have an abnormal background in waking. (**Ebersole and Pedley 2003, pp. 548–551; Blume and Kaibara 1999, pp. 156–157**)

15. (**C**): Congenital blindness is commonly associated with ipsilateral occipital spikes and slowing of occipital rhythm. Occipital abnormalities are more prominent in patients with amblyopia or blindness that started early in life. These discharges are not associated with increased risk for seizures. Other abnormalities include loss of photic response and absence of positive occipital sharp transients (POSTs) during sleep and lambda waves during wakefulness. (**Fisch 1999, pp. 340–341; Blume and Kaibara 1999, p. 352**)

16. (**B**): This EEG tracing (Fig. 5.4) is displayed in a referential average montage. There are frequent single or runs of spikes during sleep, recorded independently from the two sides. The spikes have a peak negativity (upward deflection) in the central (C3 or C4) and midtemporal regions (T3). They also have a positive end (downward deflection) in the frontal region (Fp1, Fp2, F3, F4, Fz) and vertex (Cz). These findings are typical of BECTS. Spikes could be bilateral (as in this case) or unilateral. Sleep dramatically enhances rolandic spikes, and wakefulness may abolish them completely. The frequency of rolandic spikes does not correlate with seizure frequency or severity. Most patients with BECTS will have normal EEGs in late adolescence. The background is typically normal. (**Ebersole and Pedley 2003, pp. 526–531; Blume and Kaibara 1999, pp. 159, 236–244**)

17. (**D**): Hyperventilation is routinely performed to activate generalized 3 Hz spike-and-wave discharges and absence seizures. Hyperventilation effect is more pronounced in children, resulting in high-amplitude bisynchronous delta activity. The morphology of the response can be notch shaped or sharply contoured. Only the presence of a classic generalized spike-and-wave or spike-and-slow wave morphology or clear focal abnormality is considered abnormal. (**Fisch 1999, pp. 219–223; Blume and Kaibara 1999, pp. 4–5, 89–91**)

18. (**D**): Benign occipital epilepsy is divided into early onset (Panayiotopoulos) or late onset (Gastaut) types. Clinically, they are associated with visual symptoms and vomiting with frequent secondary generalization. Most children outgrow these seizures.

Interictal EEG epileptiform discharges are seen in approximately 80% of patients, consisting of unilateral or bilateral repetitive spikes or sharp waves in the occipital region. Extraoccipital spikes are common in the Panayiotopoulos type, and can be the only abnormality in approximately one third of patients. Discharges are attenuated when eyes are open and activated when eyes are closed. Ictal EEG consists of rhythmic posterior delta or theta discharge in the Panayiotopoulos type and fast posterior rhythmic discharge in the Gastaut type. (**Wyllie, Gupta and Lachhwani 2005, pp. 382–385; Daube 2002, p. 118**)

19. (**C**): 14 and 6 positive spikes bursts occur normally during drowsiness, mostly in the posterior temporal region at a frequency rate of 14 and 6 Hz. FIRDA is a normal finding in children. Posterior slow waves of youth are delta activity superimposed to the occipital rhythm. It reacts to eye opening and closure in the same manner as the posterior rhythm. Frontal sharp waves are abnormal in children or adults. (**Daube 2002, pp. 115–118**)

20. (B): At a gestational age of 28 to 30 weeks, the EEG activity usually consists of a burst of mixed frequency (mostly delta frequency) in a discontinuous manner interspaced with periods of EEG attenuation lasting for few seconds to 1 to 2 minutes. The higher amplitude of the bursts occurs in the posterior region. This EEG pattern is referred to as *tracé discontinu*. At this age, there is no distinction between sleep and awake states. (**Mizrahi, Hrachovy and Kellaway 2003, pp. 56–59**)

21. (C): EEG voltage attenuation can result from cortical injury or dysfunction or signal attenuation by increased distance between the cortex and recording electrodes. Hemispheric voltage attenuation is commonly seen with Sturge-Weber disease. (**Ebersole and Pedley 2003, p. 316**)

22. (C): The posterior rhythm is first seen at the age of 3 months with a frequency of 3 to 4 Hz. It reaches 5 to 6 Hz around the age of 1 year and increases to 6 to 8 Hz by 5 to 6 years of age. A typical adult range alpha frequency is reached at the ages of 8 to 12. On average, two third of children of age 9 will have a mean posterior rhythm of 9 Hz. (**Daube 2002, pp. 113–114; Ebersole and Pedley 2003, pp. 102–104**)

23. (B): Neonatal seizures are not seen in patients younger than 34 to 35 weeks of conceptual age. Ictal discharges can last for 10 seconds or more and often arise from central regions, followed by temporal. In the exception of myoclonic seizures, all neonatal seizures are unifocal or multifocal at onset, frequently associated with abnormal background. (**Mizrahi, Hrachovy and Kellaway 2003, pp. 193–186; Fisch 1999, pp. 307–313**)

24. (C): This EEG tracing (Fig. 5.5) is displayed in a linked ear referential montage. There are independent 3 Hz single spike-and-wave discharges with shifting lateralized predominance over the frontocentral regions. In the context of staring spells suggesting absence seizures, these findings are most consistent with fragments of generalized epileptiform discharges. Fragments are commonly seen in generalized epilepsies during sleep. They tend to have a restricted field and tend to shift in predominance between the two sides. Patients with idiopathic generalized epilepsy and fragments will usually show typical generalized epileptiform discharges at other times during the recording. Bilateral independent frontal potential epileptogenicity is a less likely possibility. (**Abou-Khalil and Misulis 2006, pp. 131–133**)

25. (C): The EEG of a premature infant is more discontinuous (tracé alternant pattern) during sleep, and shows greater asynchrony in wakefulness. There is greater incidence of sharply contoured transients, especially during sleep. In addition, a premature infant tends to have active sleep onset rather than quiet sleep and higher percentage of active sleep. (**Mizrahi, Hrachovy and Kellaway 2003, pp. 55–58**)

26. (C): Delta brushes consist of high-amplitude 1 to 2 Hz delta wave transients superimposed on fast 8 to 20 Hz activity. These waves are predominantly seen in the posterior or rolandic regions. They are mainly seen at the ages of 32 to 36 weeks but should disappear by 42 weeks. They can be seen at any stage but more frequently during sleep. (**Mizrahi, Hrachovy and Kellaway 2003, pp. 58–60**)

27. (D): Frontal sharp transients (also known as *en couche frontales*) are diphasic sharp waves frequently seen in the frontal region of sleep. They can be symmetrical or unilateral with shifting predominance. They appear at approximately 35 weeks and may continue until 42 to 44 weeks. (**Mizrahi, Hrachovy and Kellaway 2003, pp. 58–64**)

28. (C): Rudimentary vertex waves may be present at birth in the central regions but by the age of 3 to 5 months they become well formed with a distinct morphology. They become particularly prominent by the age of 3 years. They often have high amplitudes with sharply contoured morphology that can be mistaken for abnormal epileptiform discharges. (**Daube 2002, pp. 114–115; Blume and Kaibara 1999, pp. 4–5**)

29. (B): Hyperventilation-induced slow activity in young children is often maximal over the posterior head region. In older children and adolescents, the response is usually maximal over the anterior head region. (**Daube 2002, p. 114**)

30. (A): From birth until 3 months of age, a central rhythm of 5 to 6 Hz is present. Later, it becomes better developed with a frequency of 6 to 8 Hz by 6 months of age. The central rhythm is considered to represent a precursor for the mu rhythm. (**Daube 2002, pp. 113–114**)

31. (B): The EEG during quiet sleep of a full-term neonate consists of discontinuous pattern and high-amplitude slow activity known as *tracé alternant*. This is seen as early as 33 to 36 weeks but becomes better formed and more synchronous by term age. Delta brushes and frontal sharp waves are also present during quiet sleep at this age. Tracé discontinu characterizes EEG activity of infants <30 weeks of age and quiet sleep up to 32 weeks of age. Persistence of tracé discontinu beyond 33 to 34 weeks is a sign of brain immaturity. (**Mizrahi, Hrachovy and Kellaway 2003, pp. 58–61**)

32. (B): This neonatal EEG tracing (Fig. 5.6) is displayed in a bipolar longitudinal montage.

The background consists of almost continuous low-medium mixed frequency. In a healthy full-term baby, this pattern represents the transition from tracé alternant to continuous slow wave sleep during quiet sleep. There is also the tonic activity of the chin myogram and regular heart rhythm, which are also characteristics of quiet sleep, the equivalent of NREM sleep.

Quiet sleep is the most sensitive state during which EEG abnormalities are seen and is the state that diminishes the most in sick newborns. (**Ebersole and Pedley 2003, pp. 174–185**)

33. (B): Infantile spasms are brief epileptic tonic contractions affecting infants and children as may be seen in West syndrome. The interictal correlate of infantile spasm is the chaotic hypsarrhythmia. The ictal correlate of infantile spasm is typically a sudden diffuse attenuation of this disorganized background, ranging from few seconds to approximately 1 minute. This ictal pattern is referred to as *electrodecremental pattern*. Following a spasm, the recording often becomes less disorganized for few seconds or minutes before the reappearance of the hypsarrhythmia pattern. (**Blume and Kaibara 1999, pp. 157–157, 217–220**)

34. (B): Electrodecremental response is an ictal onset pattern commonly seen in patients with infantile spasms. It is often associated with tonic or atonic seizure types. Electrographically, it consists of flattening or desynchronization of the EEG, and usually arises from an abnormal background. This electrodecremental response may last longer than the clinical seizure and the degree of attenuation may reflect the intensity of the clinical seizure. This EEG pattern can be masked by muscle artifact in the case of tonic seizures. (**Wyllie, Gupta and Lachhwani 2005, p. 178**)

35. (B): In neonatal EEGs, spikes or sharp waves are not synonymous to potential epileptogenicity as in adults especially when these epileptiform discharges are multifocal. Multifocal sharp transients are considered normal or of unknown significance when they occur randomly during neonatal EEG recording. But when frequent and focal, they often indicate a nonspecific encephalopathy that could result from many different causes. (**Fisch 1999, pp. 307–309; Mizrahi, Hrachovy and Kellaway 2003, pp. 93–104**)

36. (A): This neonatal EEG tracing (Fig. 5.7) is displayed in a longitudinal bipolar montage. There are symmetrical and synchronous delta brushes and frontal sharp waves in quiet sleep. The EEG background also shows synchrony with short intervals of discontinuity (tracé alternant). All of these neonatal EEG features are normal findings in premature infants of 36 to 40 weeks of conceptual age. Delta brushes (also known as *ripples of prematurity*) consist of 1 to 2 Hz high-amplitude waves with a superimposed beta "buzz". Frontal sharp waves (or *en couche frontales*) consist of symmetrical and synchronous sharp waves in the frontal electrodes. Delta brushes and frontal sharp waves can still be normally seen up to the age of 42 to 44 weeks (see Table 5.1). (**Mizrahi, Hrachovy and Kellaway 2003, pp. 56–61**)

TABLE 5.1 Developmental Electroencephalographic (EEG) Characteristics between 24 and 46 Weeks of Conceptual Age

Conceptual Age (wk)	Continuity/Synchrony of Background			Reactivity	Interburst Interval	Specific Waveforms and Patterns	
	Awake	Quiet sleep (non-REM sleep)	Active sleep (REM sleep)			Appearance	Disappearance
24–30	Not applicable	D/++++ (hypersynchrony)	D/++++ (hypersynchrony)	NR	6–12	Temporal theta bursts	
30–34	D/+	D/+ (tracé discontinu)	C/++	NR	5–8	Occipito-temporal delta brushes	Temporal alpha bursts
						Temporal theta bursts	
34–36	C/+++	D/+ (activité moyenne)	C/+++	R	4–6	Frontal sharp transients	Temporal alpha bursts
36–38	C/++++	D/++ (tracé alternant)	C/++++	R	2–4		Central delta brushes
38–40	C/++++	C/+++ (tracé alternant)	C/++++	R	0–2		
40–44	C/++++	C/+++ (CSWS)	C/++++	R	—	Central rhythm	Frontal sharp transients
						Asynchronous spindles	Occipital delta brushes
44–46	C/++++	C/++++ (CSWS and spindles)	C/++++	R	—	Posterior rhythm	

REM, rapid eye movement; D, discontinuous activity; ++++; total synchrony; C, continuous activity; NR, nonreactive; 0, total asynchrony; R, reactive; CSWS, continuous slow wave sleep.
(Modified from Mizrahi, Hrachovy and Kellaway 2003, p. 56, table 4.1, and Ebersole and Pedley 2003, p. 183, figure 6.23, [with permission].)

References

1. Abou-Khalil B, Misulis K. *Atlas of EEG and seizure semiology*. Elsevier Science; 2006.
2. Blume WT, Kaibara M, Young GB. *Atlas of adult electroencephalography*, 2nd ed. Lippincott Williams & Wilkins; 2002.
3. Blume WT, Kaibara M. *Atlas of pediatric eletroencephalography*, 2nd ed. Lippincott Williams & Wilkins; 1999.
4. Daube J. *Clinical neurophysiology*, 2nd ed. Oxford Press; 2002.
5. Ebersole J, Pedley T. *Current practice of clinical electroencephalograph*, 3rd ed. Lippincott Williams & Wilkins; 2003.
6. Fisch B. *Fisch and Spehlmann's EEG primer: principles of digital and analog EEG*, 3rd ed. Elsevier Science; 1999.
7. Mizrahi EM, Hrachovy RA, Kellaway P. *Atlas of neonatal electroencephalography*, 3rd ed. Lippincott Williams & Wilkins; 2003.
8. Wyllie E, Gupta A, Lachhwani DK. *The treatment of epilepsy: principles and practice*, 4th ed. Lippincott Williams & Wilkins; 2005.

CHAPTER 6

Adult Electroencephalography

QUESTIONS

1. **Subclinical rhythmic electrographic discharge of adult (SREDA) is diagnostic of:**
   A. Subclinical seizures
   B. Dementia
   C. Encephalopathy
   D. Unknown significance

2. **When awake, a 56-year-old lady with anxiety has an attenuated EEG with an 8.5-Hz bilateral occipital rhythm that can be appreciated at a sensitivity of 4 µV/mm. This finding is most likely caused by:**
   A. Drug effect
   B. Encephalopathy
   C. Huntington disease
   D. Normal variant

3. **Pure focal cortical lesions will likely produce:**
   A. Focal slow activity
   B. Focal attenuation
   C. Focal breach rhythm
   D. Generalized slow activity

4. **What is the finding in the EEG recording shown in Figure 6.1?**
   A. Bilateral epileptiform spikes
   B. Positive occipital sharp transients of sleep (POSTS)

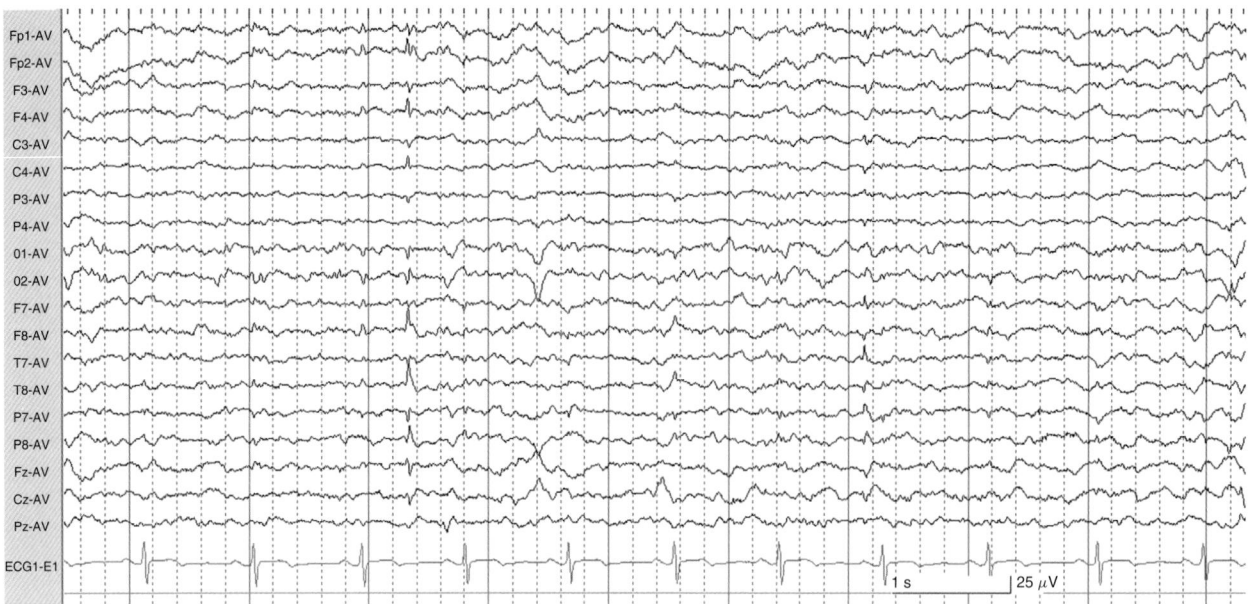

Figure 6.1.

    C. Benign sporadic sleep spikes (BSSS)
    D. Artifact

5. The EEG of a 52-year-old man reveals an occipital alpha rhythm of 8 Hz. This is considered a normal finding.
   A. True
   B. False

6. Which of the following invasive EEG techniques is not used for lateralization of the epileptogenic zone in patients with bilateral epileptiform discharges?
   A. Subdural strips
   B. Subdural grids
   C. Depth electrodes
   D. Foramen ovale electrodes

7. What is a typical ictal onset frequency for seizures presumed to start in the hippocampus?
   A. Fast alpha
   B. 5 Hz
   C. 2 to 4 Hz
   D. Generalized attenuation

8. The EEG of a 42-year-old comatose male is shown in Figure 6.2. Which of the following is true about this EEG pattern?
   A. It is specific for postanoxic brain injury
   B. It can be responsive to stimulation
   C. It can be seen in drug-induced coma
   D. It is seen only with brainstem lesions

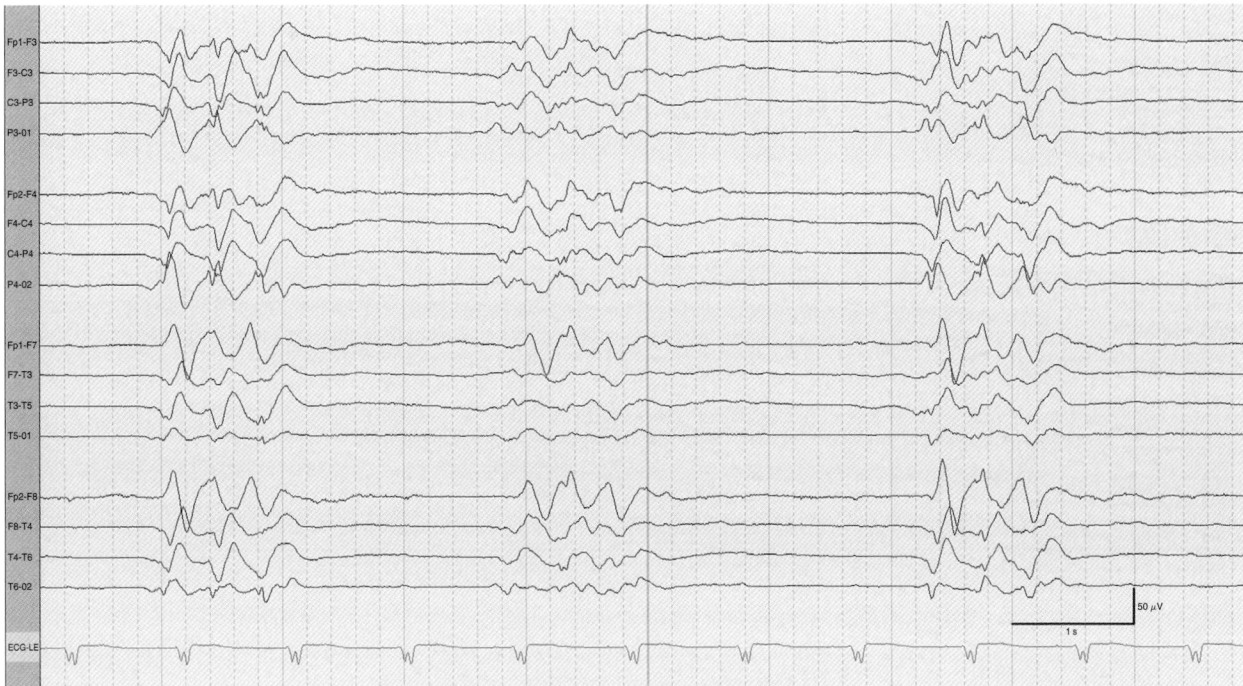

Figure 6.2.

9. During an absence seizure, generalized 3 Hz spike-and-wave discharge is faster at:
   A. Onset
   B. Mid-discharge
   C. End
   D. Does not change with evolution of the ictal discharge

10. Which of the following best describes the photomyoclonic response?
    A. Spike-and-wave activity
    B. Muscle artifact
    C. Frontal intermittent rhythmic delta activity (FIRDA)
    D. Absence of posterior rhythm

11. The EEG recording shown in Figure 6.3 is most consistent with the diagnosis of:
    A. Subdural hemorrhage
    B. Hepatic encephalopathy
    C. Subacute sclerosing panencephalitis (SSPE)
    D. Vertex waves in normal sleep

12. Temporal spikes are activated during:
    A. Photic stimulation
    B. Sleep
    C. Wakefulness
    D. Hyperventilation

13. All of the following is true about photoparoxysmal response except:
    A. Always outlasts photic stimulation
    B. Often present with spike-and-wave complexes

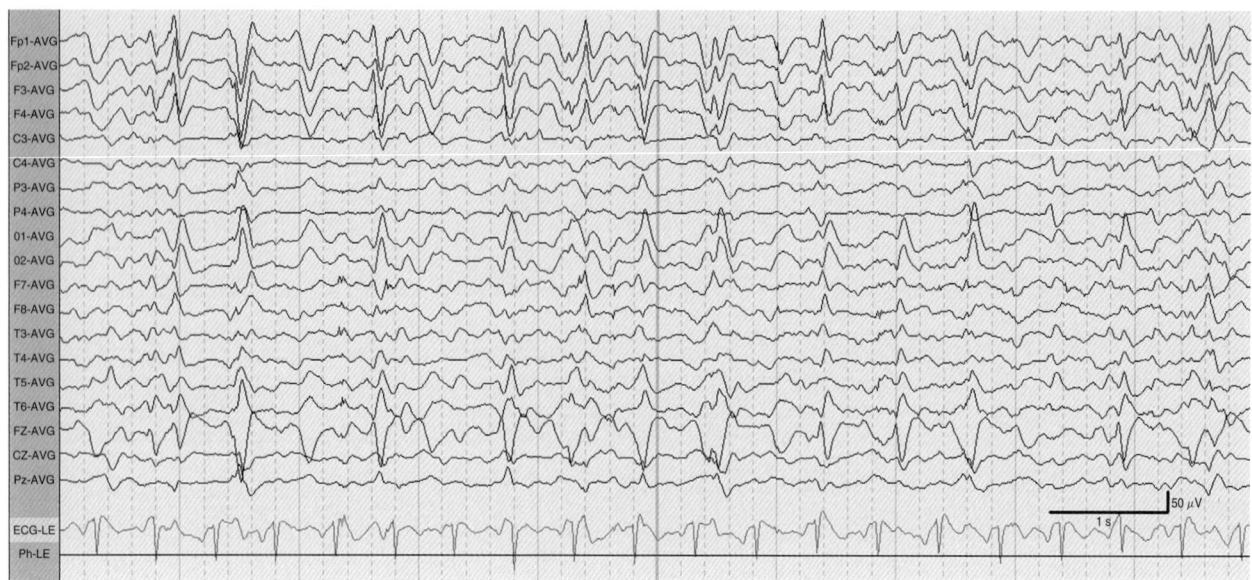

Figure 6.3.

C. Its frequency is independent of the photic frequency
D. Is often associated with generalized epilepsy

14. **During a presurgical work-up, which of the following tests involve exposure to radiation?**
    A. Positron emission tomography (PET)
    B. Single photon emission tomography (SPECT)
    C. Functional magnetic resonance imaging (fMRI)
    D. A and B

15. **The EEG recording shown in Figure 6.4 is that of a 40-year-old man during drowsiness. The EEG finding is most consistent with:**
    A. Partial seizure
    B. Trains of temporal sharp waves
    C. Normal variant
    D. Temporal intermittent rhythmic delta activity (TIRDA)

16. **During intraoperative electrocorticography, which agent can be used to activate epileptiform discharges?**
    A. Pentobarbital
    B. Methohexital
    C. Sevoflurane
    D. Propofol

17. **Hypnic jerks are associated with:**
    A. Spike-and-wave discharges
    B. K-complexes
    C. Photic stimulation
    D. Transient FIRDA

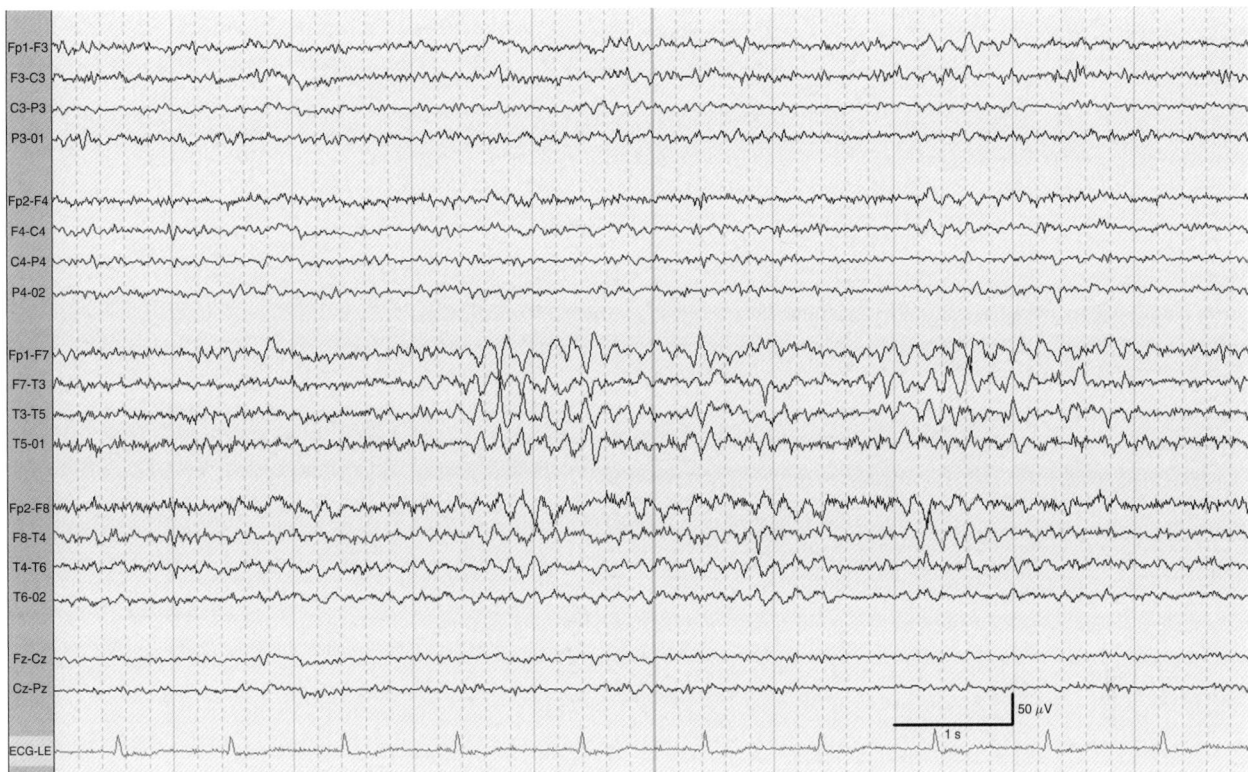

Figure 6.4.

18. Lambda waves are seen during:
    A. Heightened alertness
    B. Drowsiness
    C. Visual scanning
    D. Rapid eye movement (REM) sleep

19. All of the following is true about EEG changes during a syncopal episode except:
    A. It can produce electrocerebral inactivity
    B. The EEG normalizes instantaneously after regaining consciousness
    C. It produces generalized slow waves as consciousness is lost
    D. EEG changes are caused by reduction of blood flow to the brain

20. The EEG recording shown in Figure 6.5 is most consistent with a:
    A. Mu rhythm
    B. Wicket rhythm
    C. Gamma rhythm
    D. Lambda rhythm

21. The posterior rhythm of a 22-year-old man is of a higher amplitude (approximately 50% higher) on the right side and of a slower frequency (1.5 Hz lower) on the left side. The side that is abnormal is:
    A. Right
    B. Left

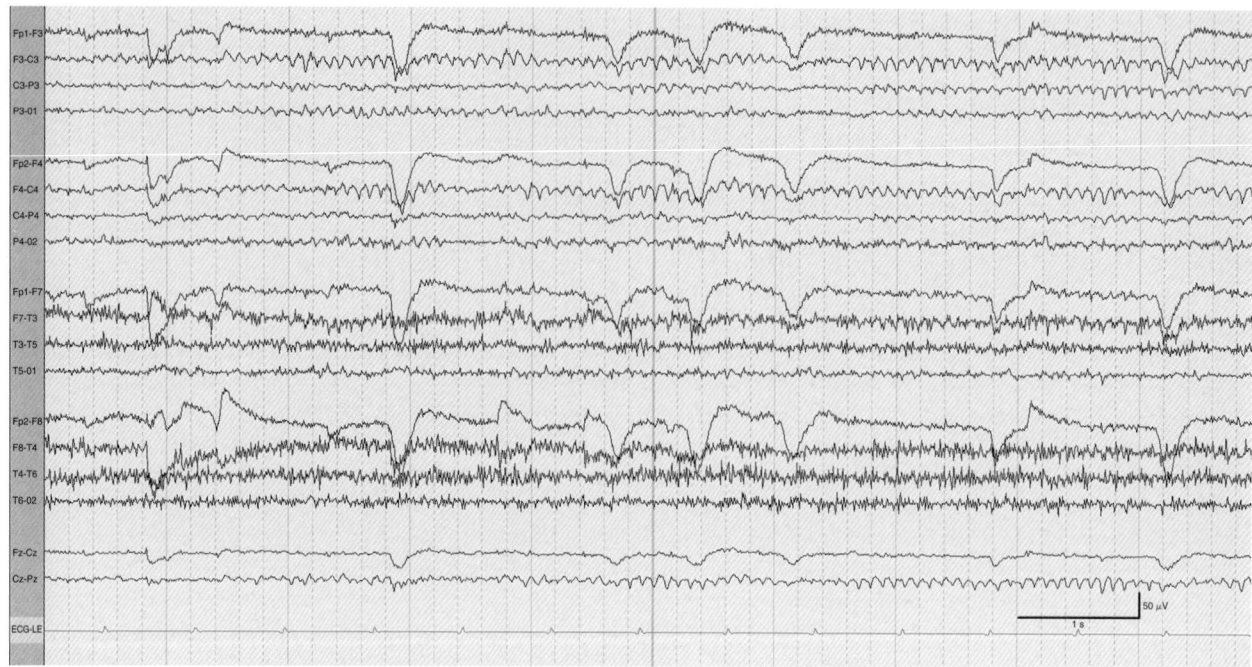

Figure 6.5.

C. Both
D. Neither

22. The EEG recording shown in Figure 6.6 is most consistent with a:
    A. Clonic seizure
    B. Tonic seizure
    C. Absence
    D. Myoclonic seizure

23. Most simple partial seizures have no scalp EEG correlates.
    A. True
    B. False

24. A 26-year-old woman has daily attacks of complete loss of consciousness. She had two previous normal routine EEGs. The outpatient procedure most likely to diagnose her attacks is:
    A. Another routine EEG
    B. 2-hour video EEG
    C. 2-hour EEG with minisphenoidal electrodes
    D. 24-hour ambulatory EEG

25. Unlike normal alpha rhythm in conscious patients, alpha coma from anoxia is:
    A. Anterior
    B. Posterior
    C. Widespread
    D. A and C

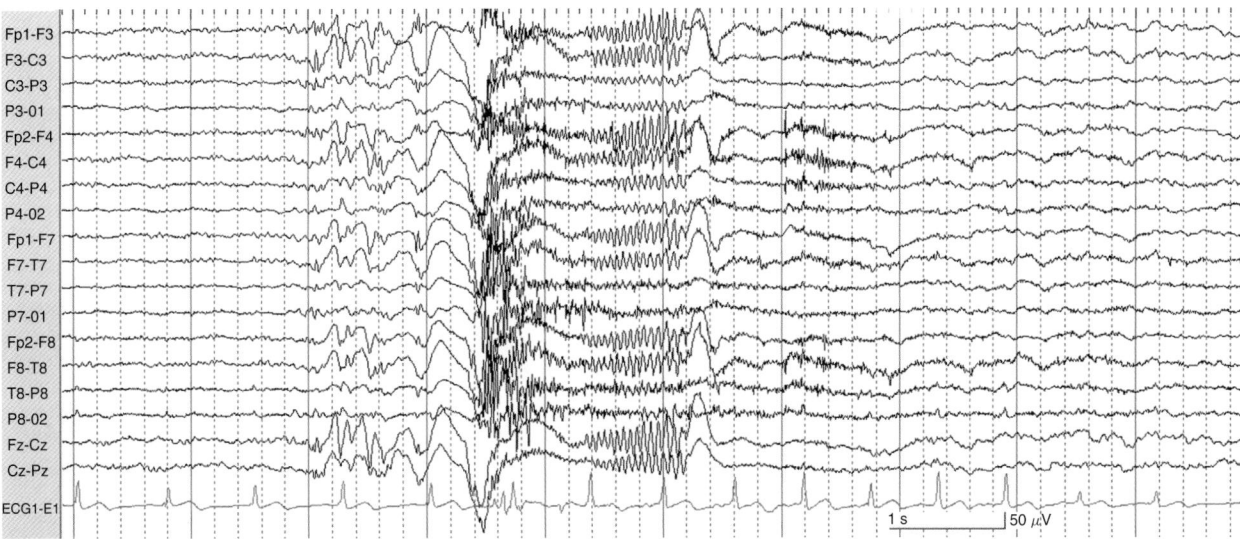

Figure 6.6.

26. Which of the following is true about periodic lateralized epileptiform discharges (PLEDs)?
    A. Herpes encephalitis is the most common etiology
    B. They can repeat at an interval of as long as 1 minute and as short as 1 second
    C. They are never ictal in nature
    D. They often indicate the presence of an acute destructive lesion

27. Which of the following is true about generalized paroxysmal fast activity?
    A. Typically seen in healthy young adults
    B. Mainly present during sleep
    C. Often associated with clonic seizures
    D. Most prominent in the posterior regions

28. Following eye closure, a brief high-frequency activity precedes a symmetrical 12-Hz posterior rhythm. This is due to:
    A. Subharmonic phenomenon
    B. Bancaud phenomenon
    C. Squeak phenomenon
    D. Breach phenomenon

29. During intracranial recording, the pattern of hippocampal ictal onset is typically:
    A. 5- to 8-Hz activity
    B. 10- to 16-Hz activity
    C. Burst of spikes
    D. Any of the above

30. In the EEG recording shown in Figure 6.7, the left temporal activity is consistent with potential epileptogenicity:
    A. True
    B. False

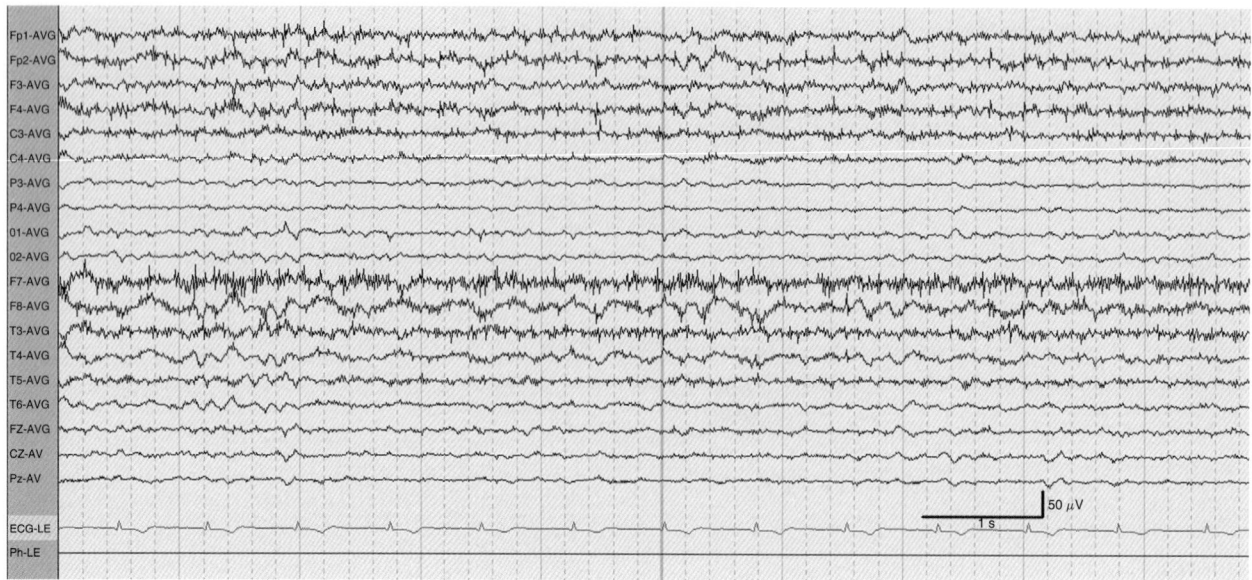

Figure 6.7.

31. The EEG recording of a 36-year-old woman with blacking out episodes is shown in Figure 6.8. This EEG shows evidence of:
   A. Left temporal breach
   B. Left temporal sharp waves
   C. Left sided sleep spindles
   D. Normal variant

32. During the relaxed state, the EEG of a 42-year-old woman shows occasional trains of notched 5-Hz activity bioccipitally. This activity attenuates with eye opening. This is most consistent with:
   A. Substance abuse
   B. Encephalopathy
   C. Normal variant
   D. Mental retardation

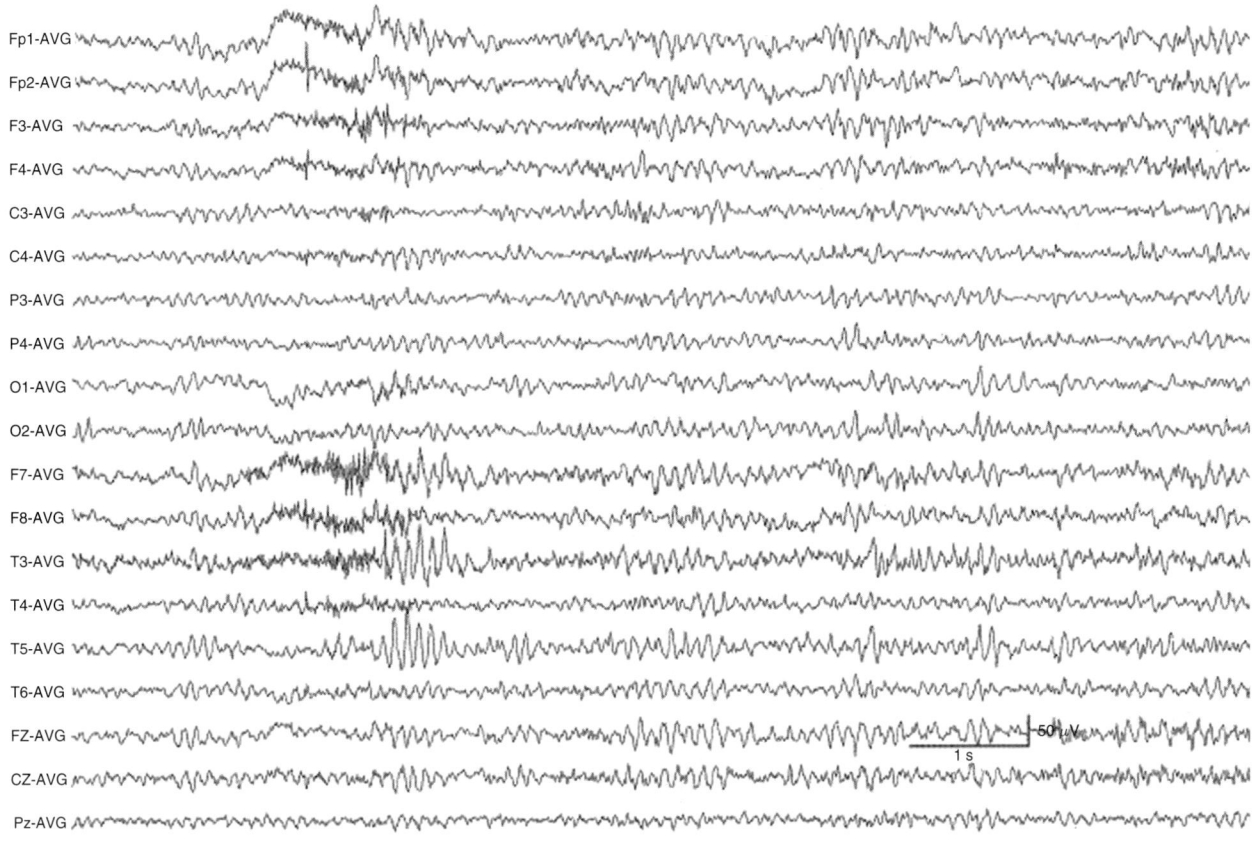

Figure 6.8.

ANSWERS

1. **(D):** SREDA is an uncommon EEG phenomenon seen mainly in older patients who are at rest or drowsy or during hyperventilation. This phenomenon starts abruptly with buildup of monophasic repetitive sharply contoured waves in the theta range (5 to 7 Hz), maximal in the parietal and posterior temporal regions, with evolution to rhythmic activity and subsequent waning. It usually involves both sides but can also be asymmetrical, lasting for an average of 40 to 80 seconds. It is not associated with clinical changes and is not associated with increased risk of epilepsy. (**Blume, Kaibara and Young 2002, pp. 112–114; Ebersole and Pedley 2003, pp. 237–238**)

2. **(D):** A low-voltage background is usually a normal variant, especially in the presence of a reactive occipital rhythm with a normal frequency. This is commonly seen in patients with acute or chronic anxiety as in this case. Other instances occur in patients performing heightened mental activity or decreased alertness to the level of drowsiness. (**Fisch 1999, p. 416**)

3. (B): Attenuation is defined as decrease in amplitude of the EEG activity. Unlike subcortical lesions that produce focal slowing of the EEG frequencies, cortical lesions produce focal decrease in amplitude or attenuation. Therefore, focal attenuation can be seen in instances of infarcts involving gray matter preferentially.

On the other hand, extra-axial lesions can also produce EEG attenuation. These include subdural hematoma, increased skull thickness, and scalp edema. In these cases, the EEG activity is attenuated due to various mechanisms, including increased distance between the cortex and recording electrode, or shunting of potential differences. (**Abou-Khalil and Misulis 2006, p. 105**)

4. (C): This EEG tracing (Fig. 6.1) is displayed in an average referential montage in light sleep.

There are two low voltage spike-like transients, one predominantly right and the other left temporal. These negative transients exhibit a smaller positivity on the opposite side. These findings are consistent with BSSS also known as *small sharp spikes* (SSS) or *benign epileptiform transients of sleep* (BETS).

They are normally seen in approximately 25% of adults during drowsiness and light sleep. They typically have a short duration (<50 ms), low voltage (<50 μV), and a single monophasic or diphasic morphology. They have a broad field maximally seen in the temporal regions, and occasionally have an oblique transverse dipole as in this case. They may have a small aftergoing slow wave.

They are distinguished from epileptiform abnormalities by their morphology, absence of background disturbance, tendency to shift between the two sides, and most importantly by their disappearance during deeper stages of sleep. (**Blume, Kaibara and Young 2006, pp. 164, 185; Ebersole and Pedley 2003, p. 241**)

5. (B): The normal frequency range of the occipital alpha rhythm in adults ranges from 8 to 13 Hz. However, the incidence of an 8-Hz occipital alpha rhythm is seen in <1% of the population. This statistically raises the suspicion of a slowed posterior rhythm. Consequently, an occipital rhythm frequency at the low end of the spectrum (such as 8 Hz) is highly suggestive of an EEG abnormality and is commonly referred to as *abnormal occipital rhythm*. (**Ebersole and Pedley 2003, pp. 102–103**)

6. (B): All of the above invasive methods except subdural grids can be used for lateralization of ictal foci, especially in suspected mesial temporal lobe epilepsy.

Foramen ovale electrodes, bilateral subtemporal strips, and depth electrodes can be used in determining lateralization of ictal foci when scalp EEG lateralization is inconclusive.

Subdural grids require a craniotomy, which should not be bilateral. Subdural grids are only used for definitive localization of ictal zone but not for lateralization. (**Ebersole and Pedley 2003, pp. 642–646**)

7. (B): Seizures presumed to start in the hippocampus commonly present with ictal onset of 5-Hz frequency involving the sphenoidal or zygomatic electrodes. The rhythmic ictal discharge appears within 30 seconds of the clinical seizure onset. This ictal pattern has a high lateralizing value (>95%) in mesial temporal lobe epilepsies. (**Ebersole and Pedley 2003, pp. 562–564**)

8. **(C):** This EEG tracing (Fig. 6.2) is displayed in a longitudinal bipolar montage.
   There are periodic one-second bursts of synchronous high amplitude theta activity separated by 2 to 3 seconds of complete EEG attenuation. These findings are referred to as *burst-suppression pattern*. This EEG pattern is nonspecific, most commonly seen in severe brain anoxia, drug-induced coma or severe hypothermia. Burst-suppression pattern is not responsive to stimulation. It has a very poor prognosis when it is a result of anoxic brain injury. Patients with this EEG pattern will be in coma. **(Ebersole and Pedley 2003, pp. 355, 408–409; Blume, Kaibara and Young 2002, pp. 487–489)**

9. **(A):** The generalized 3-Hz spike-and-wave complex associated with absence seizures is faster at the beginning of the ictal discharge. The frequency at onset is approximately 3 to 4 Hz. It slows down to 2.5 to 3 Hz by the end of the discharge. **(Wyllie, Gupta and Lachhwani 2005, p. 307)**

10. **(B):** Unlike the photoconvulsive response, the photomyoclonic response is not cerebral. It consists of contraction of frontal scalp muscles causing electromyographic (EMG) activity that is often time-locked to the light stimulus. It has no association with epilepsy. **(Abou-Khalil and Misulis 2006, pp. 67–69)**

11. **(B):** This EEG tracing (Fig. 6.3) is displayed in an average referential montage.
    There are trains of triphasic waves characterized by a small negative phase and a high voltage positive phase, followed by a lower voltage slower negative phase. Triphasic waves usually predominate in the anterior leads and frequently display an anterior–posterior phase lag of 25 to 140 ms. Triphasic waves are not specific but are commonly seen with toxic-metabolic encephalopathies, notably hepatic. **(Ebersole and Pedley 2003, pp. 354–355, 412–415)**

12. **(B):** Temporal spikes have a strong association with epilepsy. Sleep, especially non–rapid eye movement (REM) sleep, markedly activates temporal spikes and sharp waves. It is estimated that >90% of patients with temporal lobe seizures will have epileptiform discharges during sleep. **(Wyllie, Gupta and Lachhwani 2005, pp. 173–174)**

13. **(A):** The photoparoxysmal or photoconvulsive response is triggered by flashing lights during photic stimulation. It usually consists of generalized spike-and-wave discharges that may or may not outlast the light stimulation. Its frequency is usually slower than the frequency of the flashing lights. It is most often associated with generalized epilepsy. **(Abou-Khalil and Misulis 2006, p. 68)**

14. **(D):** SPECT and PET are often performed during epilepsy monitoring for localization of ictal zones in presurgical evaluations.
    SPECT is commonly performed with the use of ^{99m}Technetium-hexamethylpropylene amine oxime (^{99m}Tc-HMPAO) tracer, a nuclear isomer that emits gamma rays.
    PET uses metabolically active molecules that emit positrons, which rapidly get annihilated upon meeting electrons, emitting gamma rays. The molecule most commonly used clinically for this purpose is 18F-fluorodeoxyglucose (FDG). **(Wyllie, Gupta and Lachhwani 2005, pp. 993–1008)**

15. (C): This EEG tracing (Fig. 6.4) is displayed in a longitudinal bipolar montage.

There are bursts of rhythmic theta activity with notched appearance during drowsiness. This activity is predominant in the left mid-temporal region. This EEG pattern is known as *rhythmic midtemporal discharges* (RMTD) or *rhythmic temporal theta burst of drowsiness* (RTTBD). RMTD was also previously referred to as *psychomotor variant*. This EEG pattern can be bilateral or independent over the two hemispheres as in this case. It is a normal variant of drowsiness in adolescents and adults. Unlike ictal discharges, this EEG pattern does not interrupt the background and lacks evolution in frequency, amplitude, or distribution, which is typical of ictal events. (**Ebersole and Pedley 2003, pp. 236–238**)

16. (B): Electrocorticography is sometimes performed intraoperatively during epilepsy surgery. It records brain activity directly from the cortex. Recording of epileptiform discharges may help to delineate margins of the epileptogenic zone for resection or placement of subdural electrodes. In the absence of spontaneous epileptiform discharges, methohexital (a rapid short acting barbiturate) can be used to activate epileptiform discharges. This effect occurs only at low doses of methohexital; higher doses of this drug will suppress brain activity, similarly to other anesthetic drugs. (**Ebersole and Pedley 2003, pp. 267, 686–689**)

17. (B): Myoclonic or hypnic jerks occur during the transition from drowsiness to sleep in normal patients. It is often accompanied by vivid sensory feelings like falling off the bed. Hypnic jerks are associated with vertex waves and K-complexes.

When they become frequent, they can disturb sleep and are then classified as periodic limb movement disorders. (**Daube 2002, pp. 407–408**)

18. (C): Lambda waves are positive occipital notched waves seen during visual scanning of a complex picture. Saccadic eye movements are usually involved in the generation of these waves. Lambda waves are attenuated with eye closure. They can be asymmetrical, leading to misinterpretation as an abnormality.

Lambda waves share the same morphology as POSTS, which are exclusively seen during sleep. (**Ebersole and Pedley 2003, p. 126; Blume, Kaibara and Young 2002, pp. 88–90**)

19. (B): Syncope is commonly listed in the differential diagnosis of seizures when spells of loss of consciousness are evaluated. However, rarely dizziness or syncopal episodes are caused by seizures.

During syncope, EEG changes can range from slowing of the posterior background and generalized slow activity to electrocerebral inactivity when consciousness is fully lost. These changes depend on the degree of reduction of the cerebral blood flow. Often, the generalized slow activity slowly fades over seconds to several minutes with recovery of consciousness.

Careful review of the EEG tracing in patients with spells of loss of consciousness may reveal cardiac arrhythmias and thereby warrant the addition of cardiac telemetry. (**Fisch 1999 p. 372; Ebersole and Pedley 2003, pp. 631–634**)

20. (A): This EEG tracing (Fig. 6.5) is displayed in a longitudinal bipolar montage.

There is a transient arciform activity of alpha range frequency independently seen in the centroparietal regions. This is consistent with a mu rhythm, a normal

EEG variant. Mu rhythm is present in approximately 20% of young adults and is less common in children and the elderly. It has a typical frequency of 8 to 10 Hz and is seen in the central regions symmetrically or with shifting predominance, during wakefulness. It is considered to be a ubiquitous rhythm of the sensorimotor cortex. Mu rhythm does not block with eye opening, but blocks unilaterally with the movement of the opposite limb. (**Ebersole and Pedley 2003, pp. 108–109, Blume, Kaibara and Young 2002, pp. 64–65**)

21. (**B**): The amplitude of the alpha rhythm is variable among patients and also waxes and wanes within the same patient. The typical amplitude is 15 to 45 μV, highest in the occipital regions. Asymmetry of the amplitude is common where the right side may consistently show higher amplitudes not exceeding 50% compared to the left side. This is primarily due to the differences in bone thickness.

As for the frequency of the alpha rhythm, it normally ranges between >8 Hz and ≤13 Hz. Asymmetries >1 Hz between the two sides is considered abnormal. The side with the lower frequency is usually regarded as the abnormal side. (**Fisch 1999, pp. 185–190; Ebersole and Pedley 2003, pp. 104–105**)

22. (**B**): This EEG tracing (Fig. 6.6) is displayed in a longitudinal bipolar montage.

There is a brief generalized ictal discharge of 18 to –20-Hz medium amplitude associated with a burst of muscle artifact. This EEG pattern is referred to as *generalized paroxysmal fast activity* (GPFA). GPFA is most often associated with tonic seizures, a type of generalized seizures that are typically short in duration and commonly associated with an abnormal background.

Clinically, tonic seizures may vary from a subtle upward eye deviation to generalized stiffening of all four extremities with falling. (**Wyllie, Gupta and Lacchwani 2005, pp. 321–324**)

23. (**A**): In most cases, simple partial seizures are not associated with scalp EEG changes. This has been shown to occur in 60% to 80% of simple partial seizures. (**Abou-Khalil and Misulis 2006, pp. 185–186**)

24. (**D**): As the attacks occur daily and are described as loss of consciousness, a 24-hour ambulatory EEG is the outpatient diagnostic tool that is most likely to capture attacks. Ambulatory EEG may not be appropriate with subtle attacks where direct EEG-behavioral correlation is essential. Total loss of consciousness of organic nature should be associated with EEG changes detectable by ambulatory EEG. An ECG channel could be included in the recording. (**Abou-Khalil and Misulis 2006, pp. 184–186**)

25. (**D**): The spatial characteristic of the alpha pattern seen in postarrest unconscious patients is typically widespread with anterior predominance. In the context of a coma examination, this pattern is referred to as *alpha coma*. Another major distinction from the normal alpha rhythm is that postanoxic alpha coma does not attenuate with eye opening and typically does not react to somatosensory stimulation. (**Niedermeyer and Lopes Da Silva 1999, pp. 222–223**)

26. (**D**): Periodic lateralized epileptiform discharges (PLEDs) consist of persistent periodic sharp waves that are usually lateralized to one hemisphere. They tend to

have a recurrence rate of 0.5 to 2 cps. It is usually seen with acute, fairly large destructive lesions, most commonly cerebral infarctions, hemorrhagic or ischemic, or rapidly growing tumors. PLEDs are also seen in encephalitis, notably herpes encephalitis. This EEG pattern usually lasts no more than few weeks. PLEDs are highly associated with clinical seizures in up to 70% of the cases. Although it is generally regarded as an interictal phenomenon, this is controversial, and there are documented instances where it is ictal in nature. (**Ebersole and Pedley 2003, pp. 515–516; Blume, Kaibara and Young 2006, pp. 342–346**)

27. (B): GPFA is another defining EEG feature of Lennox-Gastaut syndrome, in addition to slow spike-and-wave discharges. It consists of runs of bilaterally synchronous 15- to 20-Hz spiky activity lasting a few seconds. It is most prominent in the frontocentral regions and almost exclusively seen during sleep. Most discharges are not correlated with clinical seizures. When they are ictal, they are typically associated with tonic seizures.(**Ebersole and Pedley 2003, pp. 548–552; Fisch 1999, pp. 291–293**)

28. (C): Following eye closure, a very brief low voltage, beta range activity is referred to as *squeak phenomenon*. This is rapidly followed by the appearance of the normal alpha rhythm frequency. Counting of the alpha rhythm frequency should be done a few seconds after eye closure in awake and relaxed patients to avoid misreading the squeak frequency for the real alpha frequency. (**Fish 1999, p. 185**)

29. (D): During intracranial EEG recording, any seizure type may have an ictal onset starting with any frequency. Hippocampal seizures may start with a burst of spikes or with a rhythmic activity. It has been reported that 10- to 16-Hz ictal onset frequency was more common in mesial temporal lobe seizures and 4- to 10-Hz or 40- to 50-Hz frequencies were more common in neocortical seizures. (**Ebersole and Pedley 2003, pp. 677–678**)

30. (A): This EEG tracing (Fig. 6.7) is displayed in an average referential montage.
There is a 1.5-Hz rhythmic activity in the left temporal region. This pattern is referred to as *temporal intermittent rhythmic delta activity* (TIRDA). TIRDA is highly associated with complex partial seizures and potential epileptogenicity in the temporal region (in this example, the left temporal region), unlike irregular temporal lobe slow activity, FIRDA or occipital intermittent rhythmic delta activity (OIRDA), all of which represent nonspecific patterns. (**Ebersole and Pedley 2003, pp. 315–323; Blume, Kaibara and Young 2002, pp. 432–436**)

31. (D): This EEG tracing (Fig. 6.8) is displayed in an average referential montage.
There are trains of medium voltage arciform waves in the left temporal region. Their morphology and distribution are consistent with wicket spikes. They can occur in singles or trains, with a temporal field. They can be bilateral and have shifting bitemporal predominance. They are considered a normal EEG variant and are more prominent during drowsiness. They are differentiated from sharp waves by the following characteristics such as occurring in rhythmic trains, innocent appearing morphology with symmetrical waveform, shifting predominance, and absence of an aftercoming slow wave. Wicket spikes are the most common normal EEG variant misread as epileptiform discharges. (**Blume, Kaibara and Young 2002, pp. 103–111**)

32. (C): In healthy patients, subharmonics of the posterior rhythm may intermix with the normal alpha rhythm, during the relaxed state. In this case, the subharmonic pattern is most likely a slow alpha variant that consists of half of the regular alpha frequency, that is, 4 to 5 Hz. Another normal alpha variant pattern is the fast alpha variant, which is usually double of the regular alpha frequency, that is, 16 to 20 Hz. These normal alpha variants can have a sinusoidal or notched shaped appearance. They are distinguished from abnormal activity by their reactivity to eye opening and closure similar to the regular alpha rhythm. **(Ebersole and Pedley, p. 237)**

References

1. Abou-Khalil B, Misulis K. *Atlas of EEG and seizure semiology*. Elsevier Science; 2006.
2. Blume WT, Kaibara M, Young GB. *Atlas of adult electroencephalography*, 2nd ed. Lippincott Williams & Wilkins; 2002.
3. Daube J. *Clinical neurophysiology*, 2nd ed. Oxford Press; 2002.
4. Ebersole J, Pedley T. *Current practice of clinical electroencephalography*, 3rd ed. Lippincott Williams & Wilkins; 2003.
5. Fisch B. *Fischand and Spehlmann's EEG primer: principles of digital and analog EEG*, 3rd ed. Elsevier Science; 1999.
6. Niedermeyer E, Lopes Da Silva F. *Electroencephalography: basic principles, clinical applications, and related fields*, 4th ed. Lippincott Williams & Wilkins; 1999.
7. Wyllie E, Gupta A, Lachhwani DK. *The treatment of epilepsy: principles and practice*, 4th ed. Lippincott Williams & Wilkins; 2005.

CHAPTER 7

Clinical Epilepsy

QUESTIONS

1. **Discontinuation of phenytoin during epilepsy monitoring is most likely to first activate:**
   A. Polymorphic delta activity
   B. Epileptiform discharges
   C. Ictal discharges
   D. Triphasic waves

2. **The electroencephalography (EEG) of a migraineur will likely reveal:**
   A. Generalized slowing
   B. Occipital spikes
   C. Enhanced photic drive
   D. All of the above

3. **The EEG recording shown in Figure 7.1 is that of a 67-year-old man with dementia. The likely diagnosis is:**
   A. Picks disease
   B. Subacute sclerosing panencephalitis
   C. Creutzfeld-Jakob disease
   D. Herpes encephalitis

4. **Prolonged slow activity after hyperventilation is most likely the result of:**
   A. Hyponatremia
   B. Hypocalcemia
   C. Hypoglycemia
   D. Hypokalemia

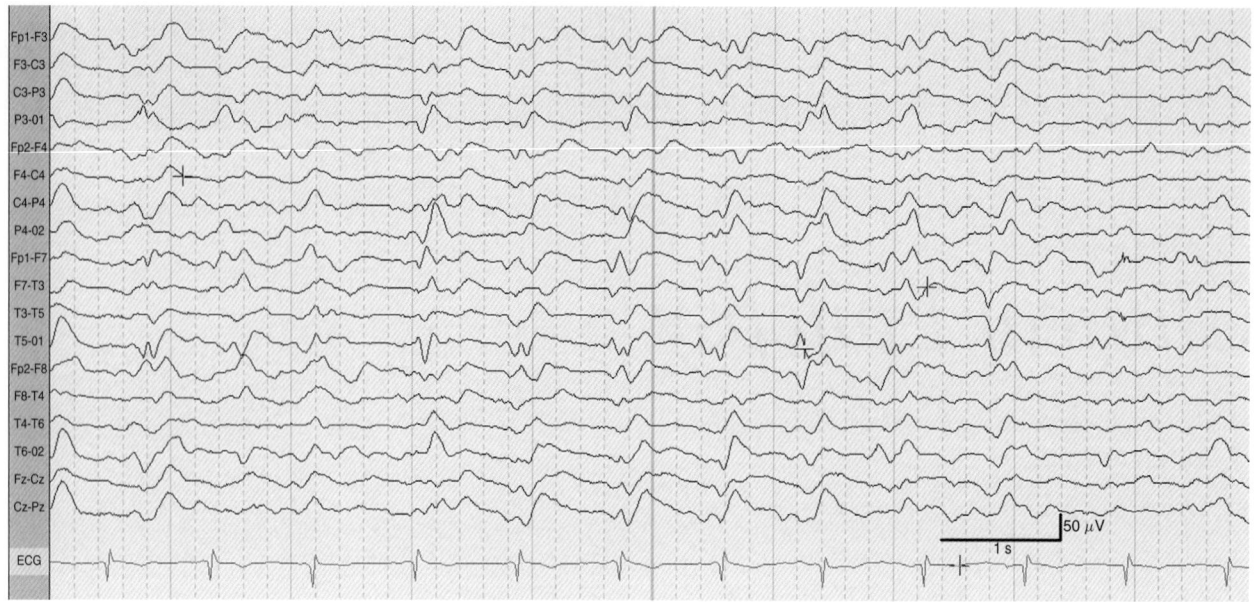

Figure 7.1.

5. A simple partial seizure with olfactory hallucination will be likely recorded at:
   A. F7
   B. T5
   C. P3
   D. F3

6. A 22-year-old man has infrequent auras of strange epigastric feeling followed by confusion, lip smacking, right arm dystonic posturing, and automatism of the left hand. In the postictal state, patient is unable to communicate for approximately 5 minutes. His seizure onset will most likely be in the:
   A. Nondominant mesial temporal lobe
   B. Dominant mesial temporal lobe
   C. Nondominant neocortical temporal lobe
   D. Dominant neocortical temporal lobe

7. Generalized 3 Hz spike-and-wave discharges were first described by:
   A. Jean-Martin Charcot
   B. Gibbs and Gibbs
   C. Henri Gastaut
   D. Hans Berger

8. The EEG recording shown in Figure 7.2 is most consistent with:
   A. Roving eye movements
   B. Sweat artifact
   C. Encephalopathy
   D. Ictal discharge

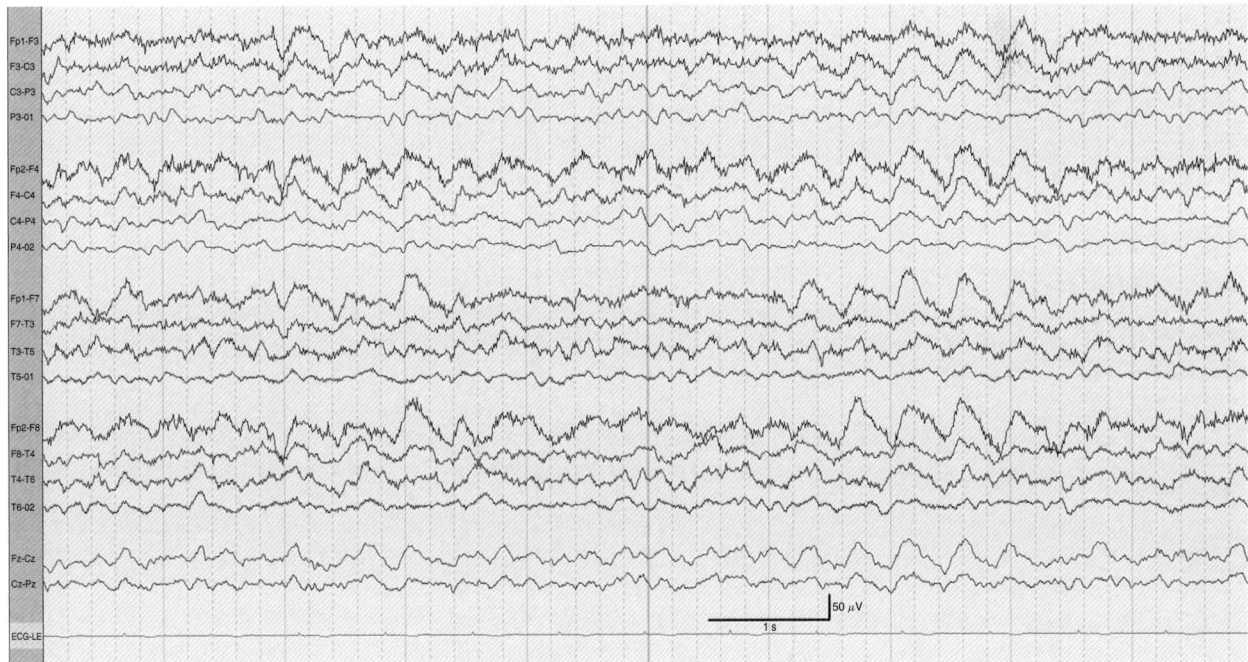

Figure 7.2.

9. Which of the following drugs can activate epileptiform discharges?
   A. Clorazepate
   B. Chloral hydrate
   C. Clozapine
   D. Promethazine

10. The EEG of a typical spell of an 11-year-old girl is shown in Figure 7.3. Among the following antiepileptic drugs, which is the most appropriate choice to make?
    A. Carbamazepine
    B. Ethosuximide
    C. Tiagabine
    D. Gabapentin

11. Seizures can be caused by all of the following electrolyte imbalances except:
    A. Hypernatremia
    B. Hypercalcemia
    C. Hypermagnesemia
    D. Hyperkalemia

12. All of the following characterize frontal lobe complex partial seizures except:
    A. Tendency to cluster
    B. Long duration
    C. Abrupt start and end
    D. Frequent secondary generalization

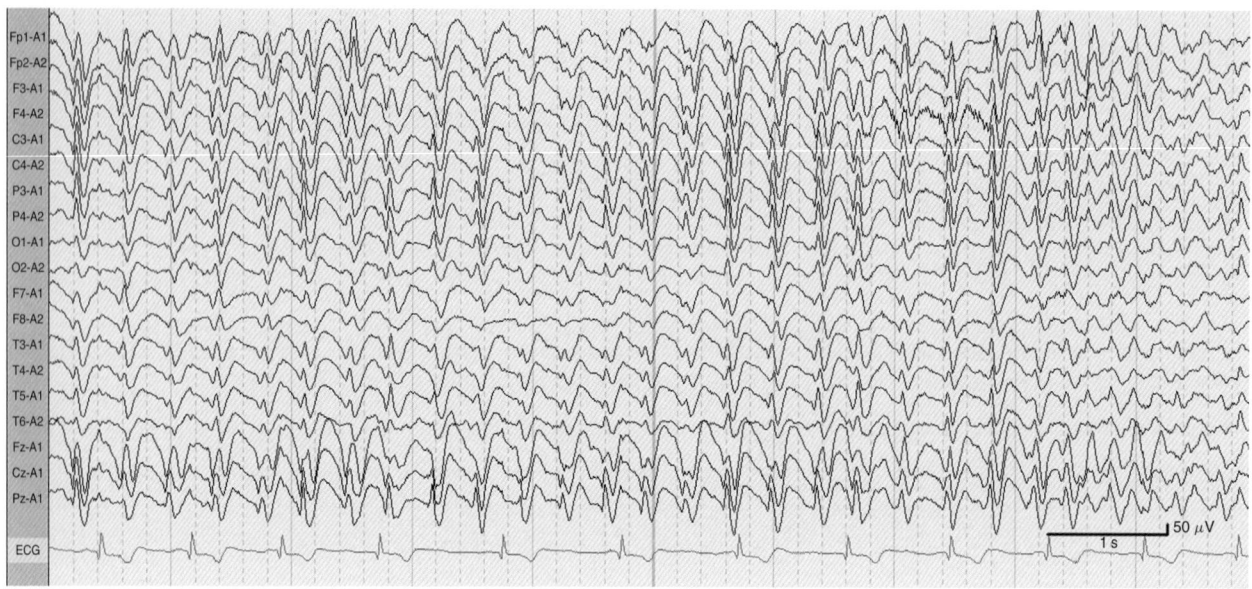

Figure 7.3.

13. The EEG recording shown in Figure 7.4 is that of a 2-year-old boy. The findings are most consistent with the diagnosis of:
    A. Lennox-Gastuat syndrome
    B. West syndrome
    C. Benign rolandic epilepsy
    D. Landau-Kleffner syndrome

14. The EEG of a 42-year-old man with an ammonia level of 82 μmol/L will most likely show:
    A. Spike-and-wave discharges
    B. Periodic lateralized epileptiform discharges
    C. Triphasic waves
    D. Normal attenuated EEG

15. In juvenile myoclonic epilepsy, myoclonic seizures are best recorded:
    A. During sleep
    B. After awakening
    C. After a heavy meal
    D. At any time

16. The EEG recording shown in Figure 7.5 is that of a 12-year-old girl. The slowing in the occipital electrodes is most consistent with:
    A. Mild encephalopathy
    B. Cortical blindness
    C. Posterior tumor
    D. Normal finding

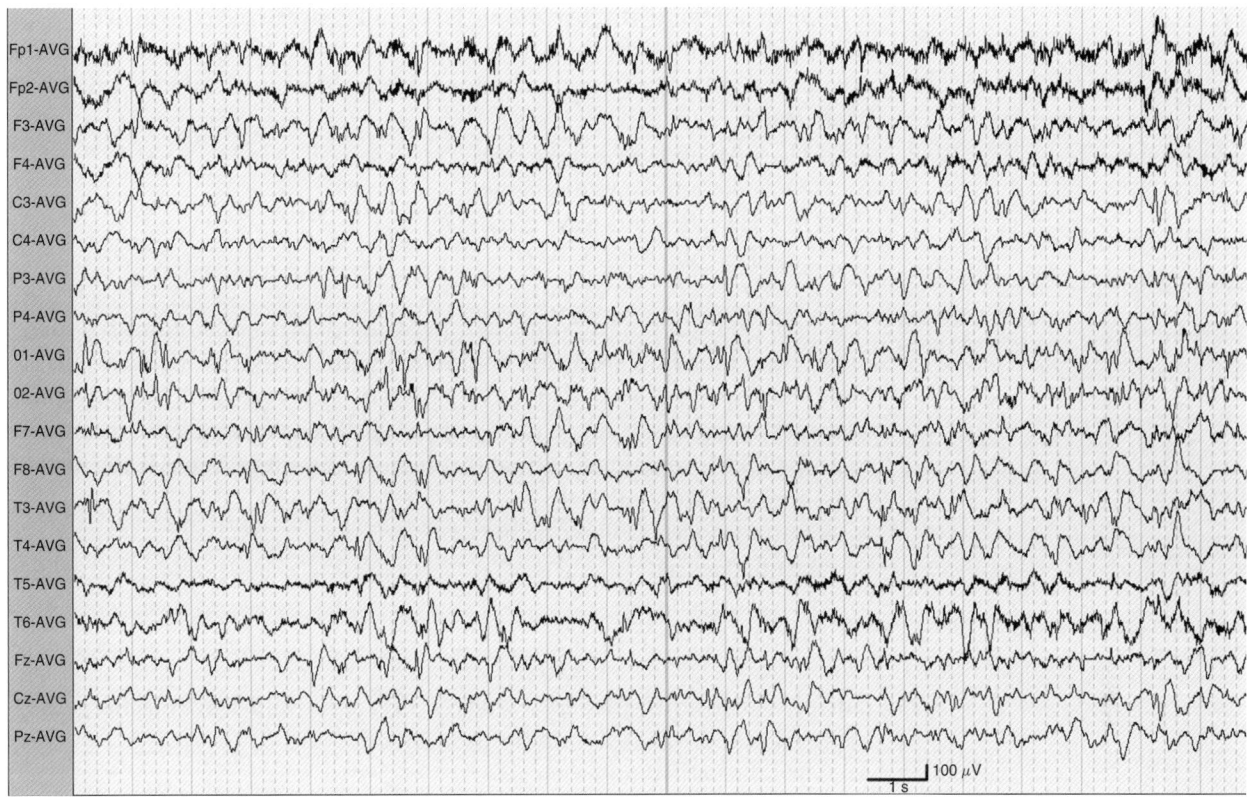

Figure 7.4.

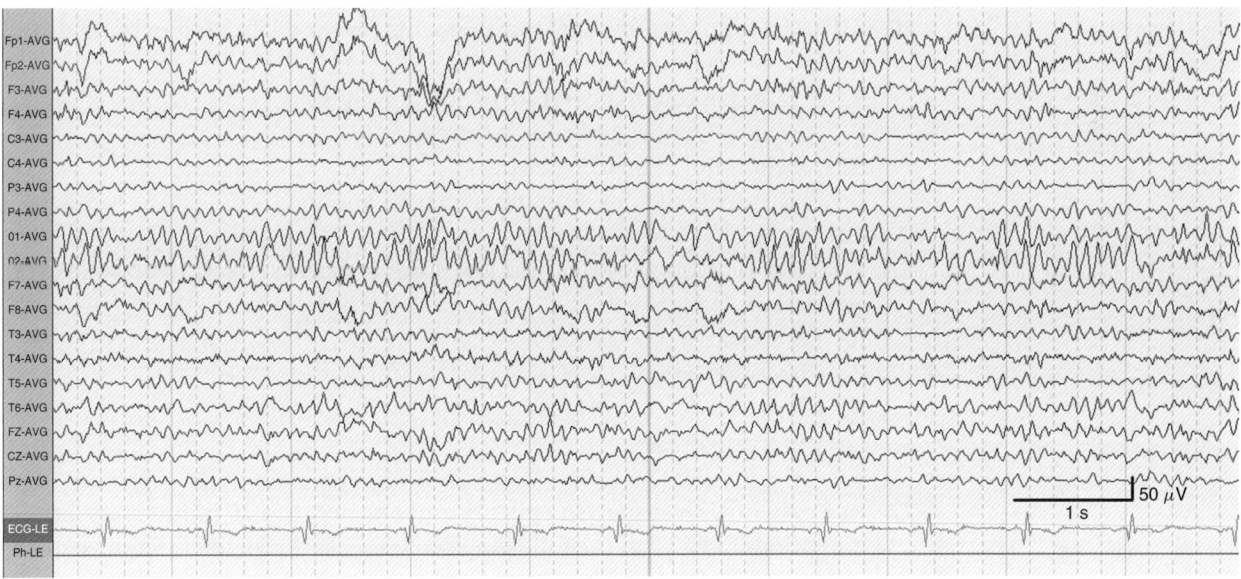

Figure 7.5.

17. A 14-year-old male with myoclonic jerks and generalized convulsive seizures is admitted for long-term video EEG study. If a generalized seizure is recorded with clonic-tonic-clonic evolution, the interictal EEG will most likely show:
    A. 1 Hz spike-and-wave discharges
    B. 3 Hz spike-and-wave discharges
    C. 4 to 6 Hz spike-and-wave discharges
    D. Temporal lobe sharp waves

18. A 26-year-old woman with unusual spells and depression is admitted for long-term video EEG study in conjunction with phenytoin withdrawal. Four short spells of asynchronous leg bicycling are recorded. Mental status was not tested during any of the recorded spells but was intact at the end of each spell. EEG remained normal throughout the study. What is the next most appropriate step?
    A. Continue video EEG recording
    B. Start fluoxetine
    C. Consult psychiatry
    D. Discharge patient home

19. During a long-term video EEG study, a 3 Hz spike-and-wave discharge is being recorded. What should the examiner do?
    A. Turn the patient to his left side
    B. Give IV lorazepam STAT
    C. Test the patient's level of consciousness
    D. Pinch the patient's hand

20. All of the following are EEG requirements for the evaluation of brain death except:
    A. A minimal core temperature of 36°C or higher
    B. A minimum of eight scalp electrodes
    C. A minimum interelectrode distance of 8 cm
    D. A sensitivity of 2 μV/mm for at least 30 minutes

21. Which of the following drugs at low doses is the least likely to activate beta activity?
    A. Lorazepam
    B. Phenobarbital
    C. Clorazepate
    D. Chloral hydrate

22. In a patient with suspected epilepsy and one negative 20 minutes awake outpatient EEG, all of the following can increase the yield of the subsequent outpatient EEG recording except:
    A. Record the next EEG for 2 hours
    B. Perform the next EEG with sleep deprivation to help obtain sleep
    C. Schedule the next routine EEG no earlier than 1 week after a typical spell
    D. Obtain a 24-hour ambulatory EEG study

23. **Which of the following conditions is the least likely to show an abnormal EEG?**
    A. Multiple sclerosis
    B. Advanced Alzheimer's dementia
    C. Glioblastoma multiforme
    D. Middle cerebral artery infarct

24. **Which EEG pattern is associated with a worse prognosis 24 hours after a cardiac arrest?**
    A. Alpha coma
    B. Spindle coma
    C. Burst suppression pattern
    D. None of the above

25. **All of the following is true about vagal nerve stimulator except:**
    A. It has shown favorable results in patients with Lennox-Gastuat syndrome
    B. It is exclusively efficacious in partial onset seizures
    C. It is used independently of patient's age
    D. It decreases the frequency of focal or generalized epileptiform discharges

26. **In patients with known partial epilepsy, a routine 20-minute EEG is likely to record epileptiform discharges in:**
    A. 25% of cases
    B. 50% of cases
    C. 75% of cases
    D. 100% of cases

27. **The ictal EEG of a seizure with vocalization and tonic posturing of the left upper extremity is most likely to be recorded at:**
    A. T3
    B. T4
    C. O1
    D. C4

28. **During video EEG monitoring, which of the following features are suggestive of nonepileptic psychogenic seizures?**
    A. Lip smacking and arm posturing
    B. Head deviation, nystagmus, and vomiting
    C. Eye closure, side-to-side motion, and asynchronous motor activity
    D. Bicycling movements and bizarre vocalization

29. **The EEG of a retired football player reveals a 7- to 8-Hz high-amplitude rhythmic sharp activity involving the T4 electrode. This finding is most consistent with the patient's history of:**
    A. Marijuana intake
    B. Past craniotomy for epidural hematoma
    C. Depression
    D. Epilepsy

30. The EEG of a 24-year-old patient with progressively worsening myoclonus reveals markedly exaggerated response to photic stimulation. Among the following, which is the most likely diagnosis?
    A. Neuronal ceroid lipofuscinosis
    B. Lafora disease
    C. Juvenile myoclonic epilepsy
    D. Dravet syndrome

31. Which of the following drugs potentiates interictal epileptiform discharges?
    A. Lithium
    B. Buspirone
    C. Fluoxetine
    D. Thioridazine

32. An EEG performed a few hours after several sessions of electroconvulsive therapy (ECT) shows right frontotemporal irregular delta activity. This likely represents:
    A. Normal post ECT finding
    B. Nonconvulsive status epilepticus
    C. Right hemispheric dysfunction
    D. Artifact

33. During epilepsy monitoring, bizarre spells with no EEG correlates are always psychogenic nonepileptic.
    A. True
    B. False

34. Which of the following is the most common stimulus type causing reflex epilepsy?
    A. Visual stimuli
    B. Reading
    C. Music
    D. Hot water

35. A 26-year-old man is taking carbamazepine for his right frontal epilepsy. The addition of which of the following drugs would likely cause him to have double vision, slurred speech, and ataxia?
    A. Erythromycin
    B. Rifampin
    C. Bupropion
    D. Antacid

36. Which of the following antiepileptic drugs does not modulate γ-aminobutyric acid (GABA)?
    A. Topiramate
    B. Tiagabine
    C. Oxcarbazepine
    D. Vigabatrin

37. A 62-year-old man is recently diagnosed with temporal lobe epilepsy. His past medical history includes kidney stones and mitral valve replacement. He is currently taking warfarin (Coumadin). Among the antiepileptic drugs listed below, which one is the least likely to interfere with his comorbid conditions and treatments?
    A. Topiramate
    B. Carbamazepine
    C. Phenytoin
    D. Levetiracetam

38. Which of the following antiepileptic drugs has the highest protein binding?
    A. Methsuximide
    B. Valproic acid
    C. Lamotrigine
    D. Oxcarbazepine

39. A 26-year-old woman had new onset of seizures with six complex partial seizures in 1 week. Which of the following is an appropriate choice?
    A. Lamotrigine
    B. Topiramate
    C. Oxcarbazepine
    D. Any of the above

40. The EEG recording shown in Figure 7.6 is most typically associated with:
    A. Hepatic encephalopathy
    B. Postanoxic brain injury
    C. Anesthesia-induced encephalopathy
    D. Severe dementia

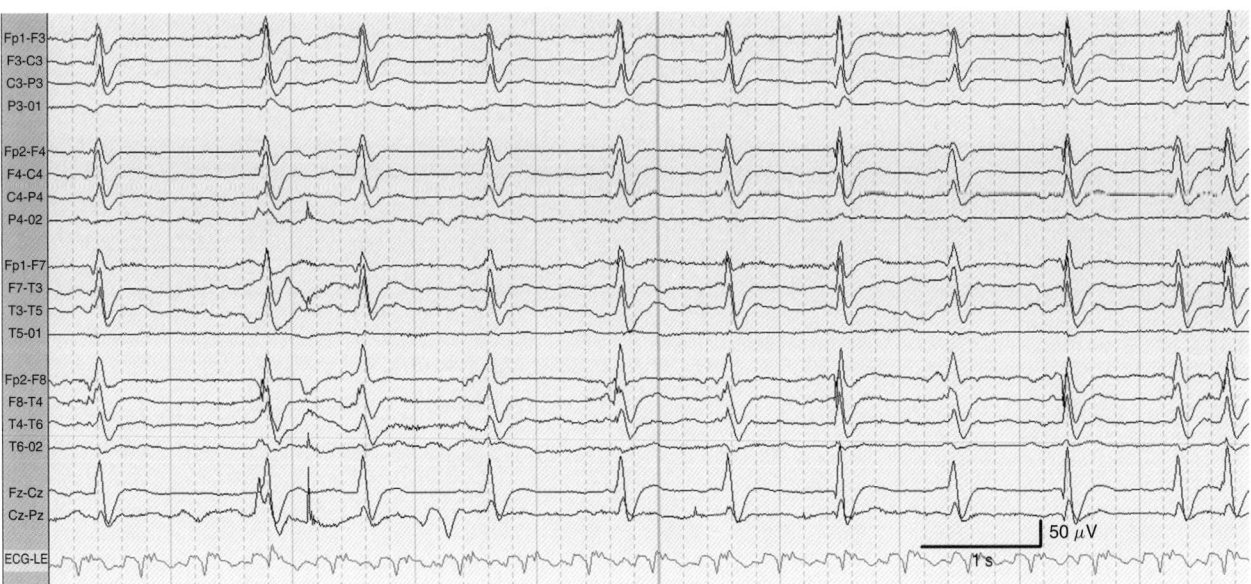

Figure 7.6.

41. The EEG recording shown in Figure 7.7 is most likely caused by:
    A. Glioblastoma multiforme
    B. Subdural hematoma
    C. Skull defect
    D. Acute pontine infarct

42. The EEG recording shown in Figure 7.8 is that of a 36-year-old woman with spells of unknown nature. The findings are most consistent with:
    A. Simple partial seizure
    B. Spindle coma
    C. Eye flutter
    D. Alpha coma

43. The EEG shown in Figure 7.9 is that of a 52-year-old woman with anxiety disorder. The EEG findings are likely due to:
    A. Diazepam
    B. Citalopram
    C. Risperidone
    D. Thorazine

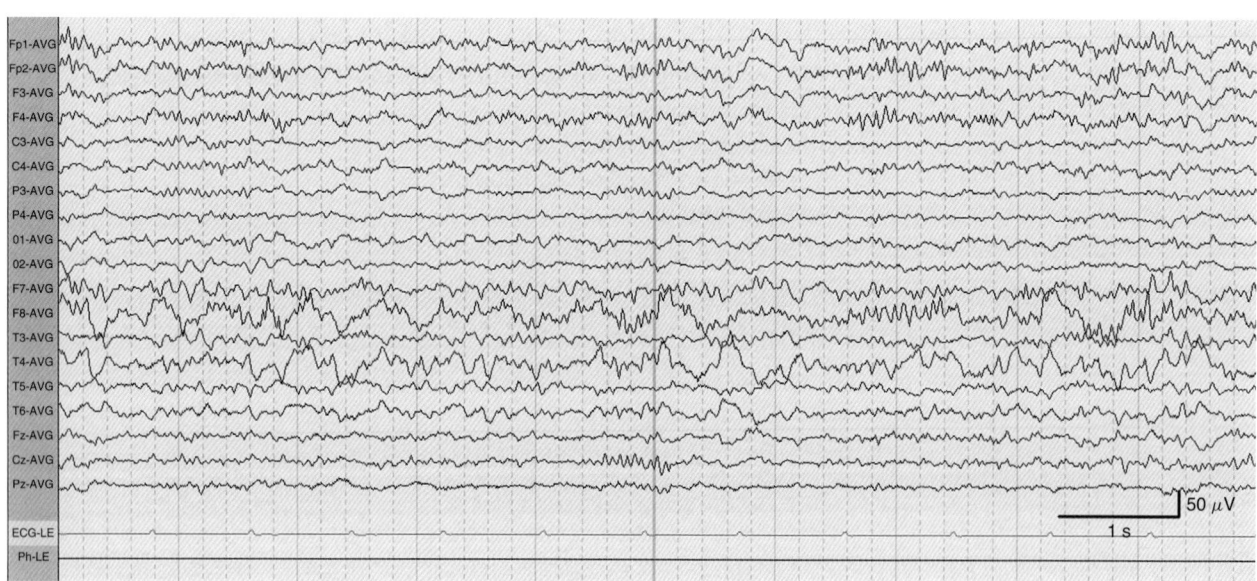

Figure 7.7.

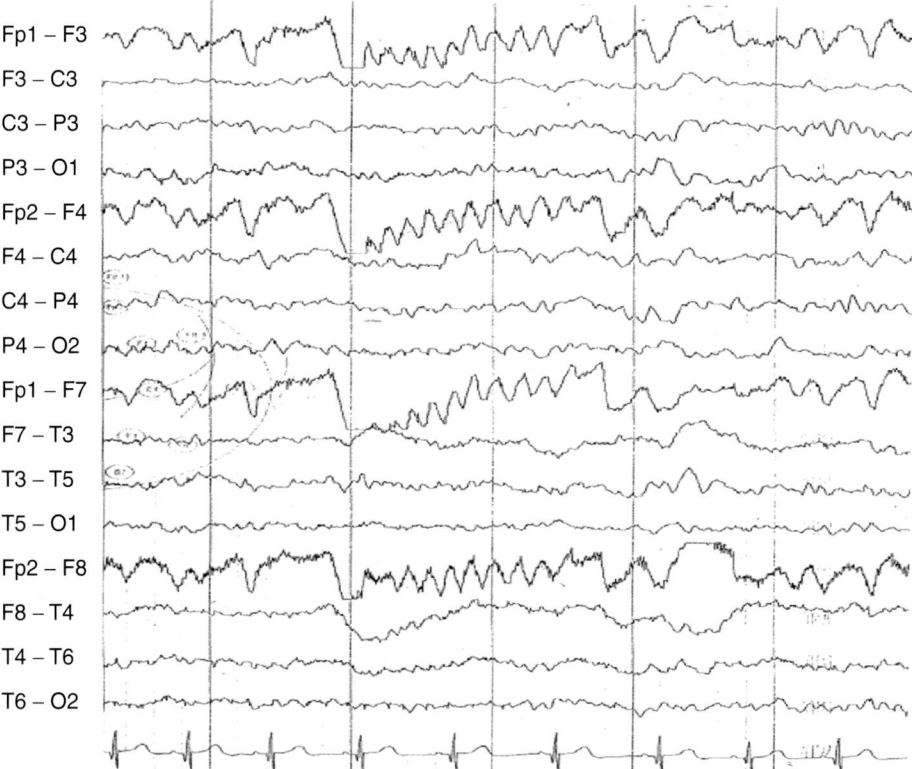

Figure 7.8.

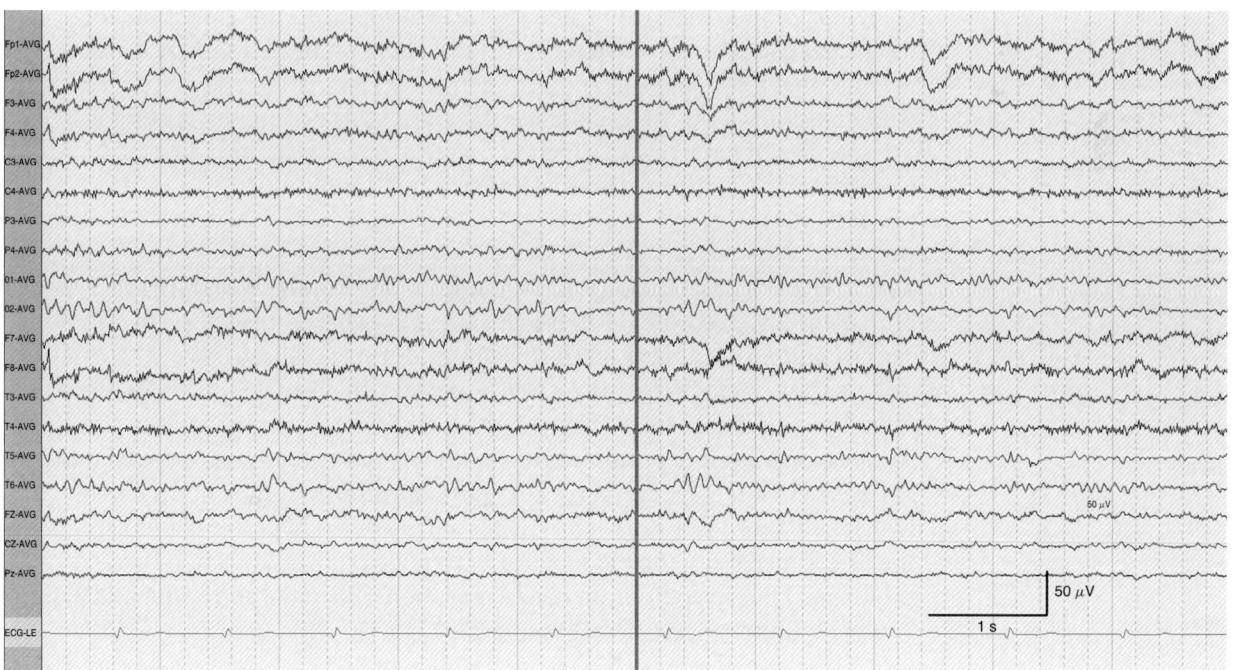

Figure 7.9.

ANSWERS

1. **(C):** Antiepileptic drugs are often discontinued during inpatient epilepsy monitoring to precipitate seizures or ictal discharges. Discontinuation of phenytoin has not been associated with activation of epileptiform discharges, but epileptiform discharges may be activated by seizures. Similarly, phenytoin therapy does not suppress epileptiform discharges. (**Ebersole and Pedley 2003, pp. 464–467**)

   On the other hand, valproate has been shown to reduce or suppress generalized 3 Hz spike-and-wave discharges.

2. **(D):** Migraine can produce a variety of EEG abnormalities. The reported EEG changes include generalized slowing (or focal in hemiplegic migraines), loss of normal alpha rhythm, enhanced photic drive, and attenuated beta activity. Occipital spikes were also reported in basilar migraines. In the absence of specific EEG changes in migraine, there is little justification in obtaining EEGs for migraineurs. (**Ebersole and Pedley 2003, p. 324**)

3. **(C):** This EEG tracing (Fig. 7.1) is displayed in a longitudinal bipolar montage. The background activity is disorganized with diffuse irregular delta activity. There are frequent periodic broad sharp waves with a left frontotemporal predominance recurring at 0.5 to 1 cps. These findings along with the clinical history of dementia are consistent with the diagnosis of Creutzfeld-Jakob disease. Creutzfeld-Jakob disease is characterized clinically by a rapidly progressing dementia, motor dysfunction, and startle myoclonus. In this condition, the discharges can be asymmetrical (in this case more on the left). These discharges may become more frequent and persistent after stimulation and maybe associated with myoclonic jerks. In the absence of clinical information, this pattern is nonspecific. Of note, this EEG pattern may be transient in the course of Creutzfeld-Jakob disease. (**Blume, Kaibara and Young 2002, pp. 462–468**)

4. **(C):** Hyperventilation is an activation technique routinely performed in outpatient EEG recording. Patients are asked to hyperventilate for 3 or 5 minutes. This technique is primarily performed to activate generalized 3 Hz spike-and-wave discharges and absence seizures. In normal individuals, there is appearance of diffuse high-voltage synchronous or asynchronous delta activity. This response is more prominent in children and becomes progressively milder and shorter with older age. It usually subsides within 1 minute of the end of hyperventilation. However, it is well documented that hyperventilation effect on the EEG depends also on blood glucose. Hypoglycemia is known to cause a prolonged EEG response from hyperventilation. (**Ebersole and Pedley 2003, pp. 130–132; Blume, Kaibara and Young 2002, pp. 81–87**)

5. **(A):** Seizures with olfactory hallucinations most often arise from the uncus, and are most likely to involve the anterior temporal (F7) and midtemporal (T3) electrodes. (**Niedermeyer and Lopes Da Silva 1999, p. 491; Abou-Khalil and Misulis 2006, pp. 10–12**)

6. (B): The clinical semiology of this complex partial seizure is highly suggestive of mesial temporal lobe epilepsy (MTLE). Auras of rising abdominal sensation and fearful feelings are common in that syndrome. Also, oroalimentary automatism (such as lip smacking), dystonic posturing, and limb automatism are more common in MTLE. As for lateralization, postictal aphasia is highly predictive of dominant hemisphere involvement, and dystonic posturing is contralateral to the ictal discharge. More commonly seen in neocortical temporal lobe epilepsy are auditory auras and early facial clonic activity. (**Wyllie, Gupta and Lachhwani 2005, pp. 365–367**)

7. (B): Gibbs and Gibbs were the first to describe the generalized 3 Hz spike-and-wave discharge seen in absence seizure. Hans Berger was the first physician who recorded brain activity from humans using electrodes. Jean-Martin Charcot was a famous 19th century neurologist. Henri Gastaut was a neurologist with extensive original work on epilepsy. (**Niedermeyer and Lopes Da Silva 1999, pp. 5–7; Wyllie, Gupta and Lachhwani 2005, pp. 169–170**)

8. (C): This EEG tracing (Fig. 7.2) is displayed in a longitudinal bipolar montage. There is a bilateral frontal intermittent rhythmic delta activity (FIRDA). This EEG pattern indicates a nonspecific generalized cerebral dysfunction. It is frequently seen in patients with encephalopathy. Its frontal predominance is due to the fact that this pattern is regarded as a projected rhythm of disrupted thalamocortical projections. FIRDA is a normal finding during drowsiness, but can be seen with several pathologic processes. Unlike temporal intermittent rhythmic delta activity (TIRDA), this EEG pattern is not usually associated with higher risk of seizures. (**Ebersole and Pedley 2003, pp. 295–296; Blume, Kaibara and Young 2002, p. 472**)

9. (C): Clozapine is a selective dopamine receptor blocker used in refractory schizophrenia. Clozapine can induce a variety of EEG changes ranging from slowing to activation of epileptiform discharges and seizures. Epileptiform discharges are usually seen in the form of bilateral spike-and-wave discharges maximal in the anterior sagittal regions. Rapid titration and high doses of clozapine increase the risks of seizures. Clorazepate produces excessive beta activity, more than chloral hydrate. (**Ebersole and Pedley 2003, pp. 468–470**)

10. (B): This EEG tracing (Fig. 7.3) is displayed in an ear referential montage. There is a continuous and rhythmic 3-Hz spike-and-wave discharge with bifrontal predominance. This EEG pattern is consistent with a typical absence seizure.

Ethosuximide is the drug of choice for isolated typical absence seizures. The remaining three antiepileptic drugs are effective against partial onset seizures only. (**Blume, Kaibara and Young 2002, pp. 328–329; Wyllie, Gupta and Lachhwani 2005, pp. 823–824**)

11. (C): Several electrolyte derangements can provoke seizures. These include hyper or hyponatremia, hyper or hypocalcemia, hypomagnesemia, or hypophosphatemia. The critical factor in precipitating seizures is the rate of electrolyte concentration changes rather its absolute concentrations. On the contrary, hypermagnesemia is known to be protective against seizures and magnesium sulfate is the drug of choice in seizure prevention in patients with eclampsia. (**Ebersole and Pedley 2003, pp. 360–361**)

12. (B): After temporal epilepsy, frontal lobe epilepsy is the second most common type of partial epilepsy accounting for approximately 20% to 30 % of cases. Frontal lobe complex partial seizures are relatively brief seizures with an abrupt start and end, with a tendency to cluster or cause status epilepticus. They also tend to rapidly generalize. Often they have absent or short postictal states. (**Wyllie, Gupta and Lachhwani 2005, pp. 367–368**)

13. (B): This EEG tracing (Fig. 7.4) is displayed in an average referential montage. There are high-voltage disorganized waves (shown in the calibration mark) intermixed with multifocal asynchronous spikes and sharp waves. There are also intervening periods of generalized attenuation. These findings are most consistent with hypsarrhythmia, an EEG pattern typically associated with infantile spasms of West syndrome. Hypsarrythmia may be diffuse or may predominate in one hemisphere. The periods of attenuation are variable and are best appreciated during non–rapid eye movement (NREM) sleep. This EEG pattern may be missed if sensitivity of the recording is not adjusted to allow better visualization of the giant waveforms. (**Blume, Kaibara and Young 2002, pp. 157–159, 212–216**)

14. (C): Hyperammonemia occurs with hepatic encephalopathy. EEG findings depend on the degree of encephalopathy and vary from slowing of the background to severe attenuation. In moderate stages of hepatic encephalopathy like in this case, triphasic waves are commonly seen in adult patients. Triphasic waves are not specific to hepatic failure, but are reported to occur more commonly in this condition than in other metabolic states. (**Ebersole and Pedley 2003, pp. 359–360**)

15. (B): Myoclonic seizures are the key seizure type in patients with JME. They consist of mild to moderate jerks of the arms, legs, or trunk. Similarly to generalized tonic–clonic seizures of JME, they tend to cluster after awakening in the morning or after naps, especially after sleep deprivation. Other known triggers are fatigue, stress, and alcohol. (**Wyllie, Gupta and Lachhwani 2005, pp. 392–393**)

16. (D): This EEG tracing (Fig. 7.5) is displayed in an average referential montage.
There is a normal symmetrical 11 Hz posterior rhythm intermixed with transient slower activity in the delta range during wakefulness. This is consistent with the posterior slow waves of youth. This EEG pattern is common in children aged 8 to 14 years but can be seen up to 21 years of age. It shares the same distribution and reactivity as the alpha rhythm. (**Ebersole and Pedley 2003, pp. 115–123; Blume, Kaibara and Young 2002, pp. 43–45**)

17. (C): A generalized seizure starting in the clonic phase is usually caused by a cluster of repeated myoclonic seizures before generalization. This is commonly seen in JME, which typically has an interictal pattern of 4 to 6 Hz spike-and-wave or polyspike-and-wave discharges. (**Wyllie, Gupta and Lachhwani 2005, pp. 393–394**)

18. (A): The stereotypy of these spells is highly suggestive of seizures. The brief duration, the bicycling movements, and the absence of a postictal state are all highly suggestive of frontal lobe seizures. The epileptic nature of the attacks may become more clear as seizures become more severe with antiepileptic drug (AED) withdrawal, particularly with secondary generalization. It is not uncommon to have absent interictal abnormalities or even no definite ictal changes on scalp EEG in some frontal lobe seizures. This is mainly due to the inability to record deeply located dipoles. (**Wyllie, Gupta and Lachhwani 2005, pp. 242–243, 249**)

19. (C): The classification of seizure types according to the International League Against Epilepsy classification is based on both clinical and electrographic features. To diagnose absence seizures, impaired awareness should be demonstrated during the generalized 3 Hz spike-and-wave discharge. (**Wyllie, Gupta and Lachhwani 2005, pp. 308–309, 347–363**)

20. (C): EEG revealing electrocerebral inactivity can be used as a supportive tool in the evaluation of brain death. The minimal requirements used are as follows:

Absence of sedation and normal core temperature
Minimum of eight scalp electrodes
Interelectrode impedance more than $>100 \, \Omega$ and $<10,000 \, \Omega$
Sensitivity of 2 μV/mm for at least 30 minutes
Appropriate filter setting
Absence of reactivity (**Daube 2002, pp. 104–105**)

21. (C): Among the drug listed, chloral hydrate is least likely to produce beta activity, especially at low doses. This is the main reason why it is commonly used as a sedative in children undergoing EEG recording. Benzodiazepines (lorazepam and clorazepate) and barbiturates (phenobarbital) are strong inducers of beta activity, even at low doses. (**Fisch 1999, p. 97**)

22. (C): Performing an additional EEG will increase the yield in patients with epilepsy. In many patients, sleep is essential to activate epileptiform discharges. Prolonging the next recording, using sleep deprivation, and obtaining a longer study with sleep are all useful. In addition, obtaining a study in the postictal state may improve the yield, because seizures activate epileptiform discharges and the EEG may also record postictal slow activity. There is no advantage in delaying the EEG after a seizure. (**Wyllie, Gupta and Lachhwani 2005, p. 170**)

23. (A): Among the neurologic conditions listed, multiple sclerosis is the least likely to show abnormalities on EEG. Middle cerebral artery infarct and gioblastoma multiforme are the most likely to show abnormalities ranging from focal slowing to epileptiform or ictal discharges because of the focal structural nature of the disease. Alzheimer's disease is almost always associated with slowing of the posterior rhythm, especially in moderate to severe cases. (**Ebersole and Pedley 2003, pp. 387–390**)

24. (C): All of the listed EEG patterns are seen in coma, caused by anoxic injury or other causes. Alpha coma consists of generalized alpha range frequency that is maximal anteriorly. Spindle coma consists of diffuse spindle-like activity. They all carry a poor prognosis but burst suppression pattern is often the pattern that immediately precedes electrocerebral silence, which is highly associated with brain death. (**Ebersole and Pedley 2003, pp. 354–359**)

25. (B): In 1997, the use of vagal nerve stimulator (VNS) was approved by the U.S. Food and Drug Administration (FDA) as an adjunctive therapy in patients with drug-resistant partial epilepsy. Studies indicated that VNS therapy is effective in partial and generalized epilepsy syndromes in adults and children. Studies have shown significant improvement in seizure control and quality of life in patients having Lennox-Gastaut syndrome. (**Wyllie, Gupta and Lachhwani 2005, pp. 969–976**)

26. (B): One of the limitations of a routine EEG study is its relative low yield in recording interictal epileptiform discharges. It is estimated that a routine EEG will record these discharges only 50% of the time. The yield increases to 90% with prolonged or repeated recording sessions and by inclusion of sleep recordings. Interictal epileptiform discharges are absent in approximately 10% of patients with epilepsy. In these cases, the diagnosis would require long-term video EEG recording to capture ictal discharges. (**Abou-Khalil and Misulis 2006, p. 175**)

27. (D): The clinical description of the seizure is consistent with a right supplementary motor seizure. Rapid onset of a brief symmetrical or more commonly asymmetric tonic posturing of one extremity is characteristic of such seizures. Often, consciousness is preserved during these seizures. Note that the EEG does not always record the ictal onset and propagation of this type of seizures. According to the international 10-20 system, C4 electrode will approximately overly the right rolandic region. (**Wyllie, Gupta and Lachhwani 2005, pp. 268–269; Abou-Khalil and Misulis 2006, pp. 9–12**)

28. (C): Video is an integral part of long-term EEG monitoring. Clinical features are always used in conjunction with EEG for characterization of epileptic or nonepileptic seizures. For the diagnosis of nonepileptic spells, several clinical features are helpful in confirming this diagnosis. Of these are eye closure, out of phase or discontinuous motor activity, forward pelvic thrusting, and others. (**Abou-Khalil and Misulis 2006, pp. 198–208**)

29. (B): In this case, a high-amplitude rhythmic spiky or sharply contoured activity confined to one electrode is most likely due to a breach rhythm. Breach rhythm represents unfiltered brain activity through a skull defect. A skull defect due to a craniotomy for hematoma evacuation is the more likely cause. The typical frequency of a breach rhythm is 6 to 11 Hz but slower or faster activity can also be seen. It is most prominent in the central regions where the mu rhythm dominates and in temporal regions where wicket spikes are common. Its amplitude may even reach threefolds the activity recorded in other regions. Breach rhythm is distinguished from abnormal epileptiform activity by its occurrence in trains, the absence of aftergoing slow waves, and lack of spread to other areas. (**Ebersole and Pedley 2003, pp. 241–243; Blume, Kaibara and Young 2002, pp. 386–389**)

30. (A): Late onset neuronal ceroid lipofuscinosis is one of the neurodegenerative diseases that is characterized by progressive dementia and myoclonus. EEG abnormalities can be seen early in the course of the disease. These include slowing of the background as well as exaggerated visual evoked potentials. The latter finding can be appreciated during photic stimulation where spiky high-amplitude photic response is seen. (**Niedermeyer and Lopes Da Silva 1999, pp. 367–368**)

31. **(A)**: Among the listed drugs, lithium has been demonstrated to activate or enhance epileptiform discharges in patients with epilepsy. Similarly, lithium can precipitate seizures in patients with known epilepsy. Reversible EEG changes mimicking Creutzfeld-Jakob syndrome findings have also been reported with the chronic use of lithium. In patients with no history of epilepsy, the potential epileptogenicity of lithium has been controversial until now. Other neuropsychiatric drugs that potentiate epileptiform discharges are clozapine, imipramine, and amitriptyline. (**Ebersole and Pedley 2003, pp. 468–473**)

32. **(A)**: ECT is used in the treatment of several psychiatric conditions, including major depressive disorder, psychosis, and catatonia. Electrical activity is transmitted to the brain in order to induce seizures while the patient is put to sleep and paralysis. ECT is known to increase slow activity for up to 1 month after several sessions. The slowing could be generalized or more commonly focal, typically affecting the temporal and rolandic areas. (**Niedermeyer and Lopes Da Silva 1999, p. 1122**)

33. **(B)**: Even in the absence of clear EEG correlates, bizarre spells should not be considered nonepileptic in nature. Complex partial seizures originating from the frontal lobes are known to cause strange behavior and may not be appreciated on scalp EEG. This is especially true for seizures originating from the cingulate gyrus or orbitofrontal region. It is not easy to distinguish these seizures from nonepileptic psychogenic spells. Their stereotyped nature and short duration can be a clue to their nature. (**Wyllie, Gupta and Lachhwani 2005, pp. 367–369; Abou-Khalil and Misulis 2006, pp. 198–200**)

34. **(A)**: Seizures of reflex epilepsy are triggered by specific stimuli. Several types of stimuli such as music, hot water, reading, eating, thinking, touch, and startle can precipitate reflex seizures; however, visual stimuli are the most common trigger for reflex seizures. This is illustrated in the case of photosensitive epilepsies where intermittent photic stimulation activates paroxysmal generalized epileptiform discharges. Other types of visual stimuli include patterns such as escalator steps, television screens, curtains, and wallpapers. Patients with photosensitive epilepsy occasionally engage in self-stimulation by waving their fingers in front of their eyes. (**Wyllie, Gupta and Lachhwani 2005, pp. 463–472**)

35. **(A)**: Carbamazepine is metabolized by the P450 complex in the liver. Erythromycin and several other drugs (including some other macrolide antibiotics, propoxyphene, verapamil, cimetidine, and fluoxetine) inhibit the metabolism of carbamazepine and cause its accumulation. The addition of the other three choices will in contrast decrease the concentration of carbamazepine. (**Wyllie, Gupta and Lachhwani 2005, pp. 665–668**)

36. **(C)**: Oxcarbazepine acts mainly through inhibition of voltage-gated sodium channels. Tiagabine is a GABA-reuptake inhibitor and vigabatrin is a selective and irreversible GABA-transaminase inhibitor. Topiramate raises brain GABA levels. (**Wyllie, Gupta and Lachhwani 2005, pp. 856, 922, 914, 907**)

37. **(D)**: All of the listed drugs are efficacious against partial onset seizures. However, one must take into consideration the comorbidites and the potential

side effects and interactions of each drug. Topiramate may precipitate more kidney stones. Phenytoin and carbamazepine may both interact with warfarin (Coumadin). Levetiracetam does not interact with coumadin and is not associated with kidney stones. (**Wyllie, Gupta and Lachhwani 2005, pp. 666–668, 885**)

38. (B): Among all antiepileptic drugs, the ones with the highest protein binding (>85%) are phenytoin, valproate, tiagabaine, and clonazepam. This is relevant when using total drug levels for clinical decision. (**Wyllie, Gupta and Lachhwani 2005, pp. 656–657**)

39. (C): The initiation of an antiepileptic drug should take into consideration the titration rates and the urgency of treatment. The high frequency of seizures seen in this patient will exclude lamotrigine or topiramate because both require a slow titration rate to reach therapeutic doses. Oxcarbazepine can be started at an effective dose. (**Wyllie, Gupta and Lachhwani 2005, pp. 735–742**)

40. (B): This EEG tracing (Fig. 7.6) is displayed in a longitudinal bipolar montage. There are generalized periodic sharp waves with a repetition rate of 1 to 1.5 cps on a diffusely attenuated background. This EEG pattern is referred to as *generalized periodic epileptiform discharges* (*GPEDs*). This EEG pattern is nonspecific but most commonly reported in comatose patients following acute, severe brain anoxia. GPEDs are commonly associated with myoclonic jerks. Similarly to postanoxic burst suppression EEG pattern, GPEDs are associated with a very poor prognosis after anoxic brain injury. (**Ebersole and Pedley 2003, pp. 408–410**)

41. (A): This EEG tracing (Fig. 7.7) is displayed in an average referential montage. There is an irregular delta activity at T4 and F8 electrodes that is variable in duration, amplitude, frequency, and morphology. This EEG pattern is referred to as *focal polymorphic delta activity* and is typically seen with focal acute or rapidly expanding structural lesions, in this case a malignant tumor. The generation of polymorphic delta activity involves white matter involvement with deafferentation of the cortex. Concomitant involvement of the cortical region is also very common, causing attenuation of this activity in the center. (**Fisch 1999, pp. 352–353; Ebersole and Pedley 2003, pp. 294–295, 515**)

42. (C): This EEG tracing (Fig. 7.8) is displayed in a longitudinal bipolar montage. There is a 2-second rhythmic activity that is confined to the frontopolar electrodes (Fp1 and Fp2) following an eye blink. This is most consistent with eye flutter. Eye flutter can produce a rhythmic activity of up to 10 Hz frequency. (**Ebersole and Pedley 2003, p. 274; Blume, Kaibara and Young 2002, p. 19**)

Eyelid flutter, like any eye movement, can be confirmed by adding infraorbital electrodes below the eye. The polarity in these electrodes will be opposite that in the frontopolar electrodes.

Ictal discharges will usually have a wider field and will evolve in amplitude or frequency. Spindle coma consists of a diffuse 12- to 14-Hz spindle-like activity in a comatose patient. Alpha coma consists of diffuse alpha range frequency in a comatose patient.

43. (A): This EEG tracing (Fig. 7.9) is displayed in an average referential montage. There is excessive diffuse beta activity, maximally in the midline and frontal regions. This is most commonly due to sedative drug effects, notably benzodiazepines (in this case diazepam) and barbiturates. Electromyogram (EMG) activity from scalp muscle contraction can display similar findings but in a different spatial distribution. Scalp muscle artifact would be most prominent in the frontal and temporal regions and least prominent in the midline regions due to absence of scalp muscles. (**Abou-Khalil and Misulis 2006, p. 110**)

References

1. Abou-Khalil B, Misulis K. *Atlas of EEG and seizure semiology*. Elsevier Science; 2006.
2. Blume WT, Kaibara M, Young GB. *Atlas of adult electroencephalography*, 2nd ed. Lippincott Williams & Wilkins; 2002.
3. Daube J. *Clinical neurophysiology*, 2nd ed. Oxford Press; 2002.
4. Ebersole J, Pedley T. *Current practice of clinical electroencephalograph*, 3rd ed. Lippincott Williams & Wilkins; 2003.
5. Fisch P. *Fisch and Spehlmann's EEG primer: principles of digital and analog EEG*, 3rd ed. Elsevier Science; 1999.
6. Niedermeyer Ernst, Lopes Da Silva F. *Electroencephalography: basic principles, clinical applications, and related fields*, 4th ed. Lippincott Williams & Wilkins; 1999.
7. Wyllie E, Gupta A, Lachhwani DK. *The treatment of epilepsy: principles and practice*, 4th ed. Lippincott Williams & Wilkins; 2005.

# SECTION 3

Evoked Potentials

CHAPTER 8

Visual Evoked Potentials

QUESTIONS

1. **The pattern shift visual evoked potential (PSVEP) equipment is set up such that the subject has to be seated from monitor at:**
   A. 1 ft
   B. 2 ft
   C. 1 m
   D. 2 m

2. **Increased papillary diameter results in:**
   A. No change in P100 latency
   B. Delay in P100 latency
   C. Decrease in P100 latency
   D. Any of the above

3. **Pattern shift visual evoked potentials are best recorded at an amplification of:**
   A. 150
   B. 1,000
   C. 2,500
   D. 20,000

4. **P100 amplitude is the highest in most healthy subjects at:**
   A. Inion
   B. Fz
   C. Vertex
   D. Pz

5. **P100 with two distinct peaks (W or bifid pattern) may be corrected by:**
   A. Use of lower field stimulation
   B. Use of upper field stimulation
   C. Use of full field stimulation
   D. Use of pattern evoked retinogram (PERG)

6. **If the visual acuity is normal:**
   A. Smaller check size produces larger amplitude P100
   B. Smaller check size produces low-amplitude P100
   C. Check size has no effect on P100 amplitude
   D. Smaller check size results in bifid P100

7. **An abnormal PERG raises the possibility of:**
   A. Low-visual acuity
   B. Inadequate fixation
   C. Optic neuritis
   D. Optic nerve tumor

8. **P100 latency in women compared to men is usually:**
   A. Prolonged
   B. Shorter
   C. Equal
   D. Any of the above

9. **Conduction defects in optic nerve, secondary to demyelination commonly produce latency delays without alteration in P100 waveform configuration.**
   A. True
   B. False

10. **Interocular P100 latency difference is not a good indicator of optic nerve dysfunction.**
    A. False
    B. True

11. **Compression of anterior visual pathways tends to produce alteration of P100 waveform with loss of amplitude but less delay in latency.**
    A. False
    B. True

ANSWERS

1. (C): During PSVEP, the patient is seated at a distance of 1 m from the monitor. Greater distance can affect the visual acuity and delay in P100 latency. The eyes are always tested one at a time and the opposite eye is covered with a patch. (**Chiappa 1997, p. 45**)

2. (C): Pattern luminance is very important in PSVEP testing and changes in it have marked effect on P100 latency. For this particular reason, normal values obtained and used for clinical interpretation in one laboratory cannot be used in another laboratory. Changes in luminance have a marked effect on P100 latency. Increased papillary diameter will illuminate more of retina resulting in false decrease in P100 latency, while decreasing the luminance will delay P100 latency. (**Chiappa 1997, p. 52**)

3. (D): Signal-to-noise ratio for PSVEP is very low. The PSVEP signal has to be amplified to make it prominent compared to the electroencephalographic (EEG) signal. Averaging increases the signal-to-noise ratio and improves signal quality. At least, amplification of 20,000 is needed to make the signal more prominent than the EEG signal. (**Chiappa 1997, p. 13**)

4. (A): In healthy subject, the P100 amplitude is highest at inion. In abnormal cases, P100 may be displaced and appear higher in amplitude at Pz. (**Chiappa 1997, p. 48**)

5. (A): P100, sometimes presents as two peaks separated by 10 to 50 ms. Bifid P100 is very rarely seen in healthy individuals. It may be produced with upper visual field contributing inverted-polarity activity to the inion resulting in bifid waveform, or because of visual field defects the inverted-polarity activity may be recorded over the contralateral scalp. This results in two peaks. Using only the lower field stimulation may correct the waveform in the first scenario or recording from laterally placed electrodes will show the true P100. (**Chiappa 1997, p. 102**)

6. (A): Smaller checks produce larger amplitude P100. However, if the checks size is too small such that the visual acuity is affected then this relationship does not hold true and P100 amplitude may decrease. This is why it is important to use check size that is not smaller than the visual acuity for the eye. (**Chiappa 1997, p. 62**)

7. (B): Inadequate fixation can result in complete absence of PERG response, while a normal PERG suggests that fixation was adequate. PERG is produced by retina with macula contributing for its generation. Inadequate fixation will result in no response. The other options will affect P100 but not PERG. (**Chiappa 1997, p. 76**)

8. (B): P100 latency is slightly shorter in women. Possible mechanisms are high deep-body temperature, brain volume, and head size. (**Chiappa 1997, p. 78**)

9. (A): Conduction defects in optic nerve secondary to demyelination most commonly produce latency delays without alteration in P100 waveform configuration. The relative preservation of P100 shape in partial optic nerve demyelination is due to inhibition of the late arriving impulses conducted slowly through demyelinated segments of the optic pathways. P100 latency reflects the function in the fastest conducting optic nerve axons. (**Chiappa 1997, p. 100**)

10. (A): Interocular P100 latency difference is probably the most sensitive indicator of optic nerve dysfunction in PSVEP. The most sensitive abnormality on visual evoked potential (VEP) with optic neuritis is interside P100 amplitude difference. However, in clinical practice the most useful abnormality to be detected on VEP with optic neuritis is interside P100 latency difference. **(Chiappa 1997, p. 102)**

11. (B): Compression of anterior visual pathways tends to produce decreased amplitude and distortion of P100 waveform. Delay in latency is not seen as much as that in demyelinating lesions, which typically delay the P100 latency. **(Chiappa 1997, p. 115)**

References

1. Chiappa KH. *Evoked potentials in clinical medicine*, 3rd ed. Lippincott-Raven; 1997.

CHAPTER 9

Brainstem Auditory Evoked Potentials

QUESTIONS

1. **Wave I is generated by:**
   A. Distal 8th nerve
   B. Inferior colliculus
   C. Lower pons
   D. Mid brain

2. **Select the correct statement:**
   A. Infants are best tested with a single phase click with rarefaction
   B. Infant responses are larger in amplitude
   C. Hearing threshold in infants is usually 50 dB NHL
   D. Evoked potential (EP) abnormalities are permanent and serial testing is not needed

3. **Wave V shown in Figure 9.1 is generated by:**
   A. Distal 8th nerve
   B. Superior colliculus
   C. Lower pons
   D. Midbrain

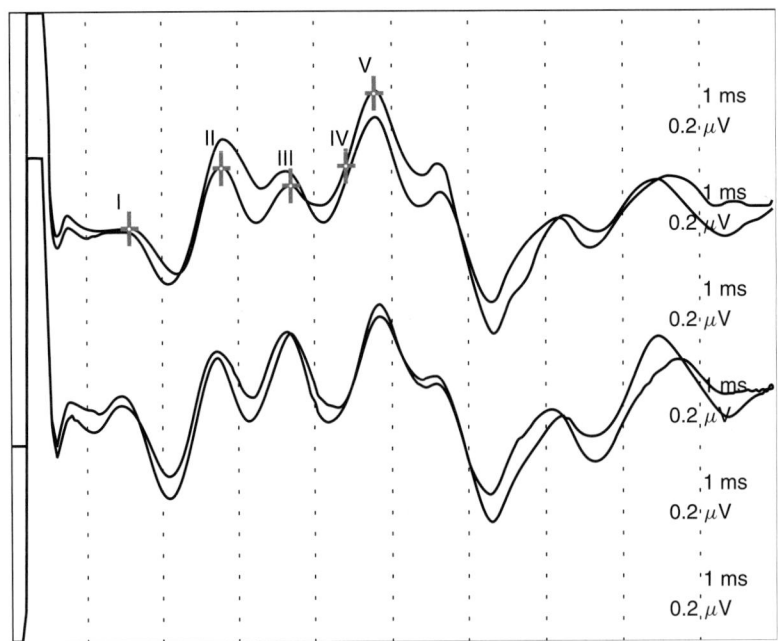

Figure 9.1. Brainstem Auditory Evoked Potential (BAEP)

4. **Click threshold is the intensity at which the subject can:**
   A. Barely hear the clicks
   B. Hear the clicks the best
   C. Above which clicks disappear completely
   D. None of the above

5. **Masking noise is:**
   A. Clicking noise used for testing the ear
   B. Low-intensity noise used for masking the contralateral ear
   C. Clicking noise producing the best waveforms
   D. Not used in humans

6. **No significant difference in latencies or amplitudes when brainstem auditory evoked potentials (BAEPs) are recorded simultaneously with needle and surface electrodes.**
   A. True
   B. False

7. **To make wave I more prominent and to get rid of cochlear microphonics:**
   A. Use contralateral ear recording
   B. Use rarefaction clicks
   C. Use condensation clicks
   D. Alternate between rarefaction and condensation

8. **For BAEP, low frequency filter should be:**
   A. 1 Hz
   B. 10 Hz
   C. 100 Hz
   D. 1,000 Hz

9. For BAEP, high frequency filter should be:
   A. 3 Hz
   B. 30 Hz
   C. 300 Hz
   D. 3,000 Hz

10. For BAEP, sweep duration should be:
    A. 10 seconds
    B. 1 second
    C. 10 ms
    D. 1 ms

11. Wave I amplitude tends to be greater with:
    A. Rarefaction clicks
    B. Condensation clicks
    C. Rarefaction and condensation clicks
    D. Not affected by click types

12. Which statement is true about wave V?
    A. Recording from contralateral ear shows highest amplitude of wave V
    B. Wave V is the most prominent peak after 1.4 ms
    C. With progressive decrease of stimulus intensity, wave V is the last peak visible in the area.
    D. Wave V is significantly delayed in cortical deafness

13. Which statement is true about wave I?
    A. Recording from contralateral ear shows wave I
    B. Wave I appears after 2.5 ms
    C. Wave I appears after 1.4 ms
    D. Wave I reverses polarity with reversal of the click polarity

14. Which statement is false about wave III?
    A. Wave III appear approximately equidistant between wave I and wave V
    B. Wave III may be markedly accentuated in the contralateral recordings
    C. It is the last wave present in this area with decreasing click intensities
    D. Wave III may be markedly attenuated in the contralateral recordings

15. Choose the correct statement:
    A. Increasing click rate results in increased absolute latencies of all BAEP waves
    B. Wave V is negative at both vertex and ipsilateral earlobe sites
    C. Waveforms in infants are often lower in amplitude than in adults
    D. BAEP adult configuration is reached by 2 yrs of age

16. In infant BAEP, sweep duration should be:
    A. <10 ms
    B. <15 ms
    C. >15 ms
    D. >20 ms

17. **Wave III is generated by:**
    A. Distal 8th nerve
    B. Inferior colliculus
    C. Lower pons
    D. Midbrain

ANSWERS

1. (A): Wave I of BAEP is a negative wave recorded at the ear, with an average latency of 1.4 ms in healthy subjects. Wave I is generated by distal 8th nerve action potentials. (**Chiappa 1997, p. 200**)

2. (A): The BAEP is recorded in normal infants at 30 dB NHL with rarefaction clicks. Infant BAEP responses are lower in amplitude. The BAEP abnormalities are usually transient and may need serial testing. (**Chiappa 1997, pp. 270, 277**)

3. (D): Wave V of the brainstem auditory evoked potential is thought to be generated at the level of the inferior colliculus and an absence of wave V would suggest a lesion of the lower midbrain at the level of the inferior colliculus. (**Chiappa 1997, p. 202**)

4. (A): Hearing threshold is the intensity at which the subject can barely hear the click. The stimulus intensity used is 65 or 70 dB above the hearing threshold. (**Chiappa 1997, p. 160**)

5. (B): If there is asymmetric gross abnormalities of hearing function then cross-stimulation may be confusing factor. A masking noise at an intensity of 30 to 40 dB less than the stimulus intensity is used in the contralateral ear. (**Chiappa 1997, p. 171**)

6. (A): Needle electrodes are as good as scalp electrodes. In order to make wave I more well defined a needle electrode is used in the anterior wall of external auditory canal. A conventional small thin wire, sterile electroencephalogram (EEG) needle electrode is used. The electrode impedance is usually approximately 5,000 ohms. (**Chiappa 1997, p. 162**)

7. (D): Rarefaction tends to have a higher amplitude of wave I. Cochlear microphonics unlike wave I change polarity with changing click polarity. Alternating between rarefaction and condensation helps in getting rid of cochlear microphonics. (**Chiappa 1997, p.171**)

8. (C): The filters are set at 50 to 150 Hz low-frequency filter and 3,000 Hz high-frequency filter with sweep duration of 10 ms. (**Chiappa 1997, p. 164**)

9. (D): The filters are set at 50 to 150 Hz low-frequency filter and 3,000 Hz high-frequency filter with a sweep duration of 10 ms. (**Chiappa 1997, p. 164**)

10. **(C):** The filters are set at 50 to 150 Hz low-frequency filter and 3,000 Hz high-frequency filter with a sweep duration of 10 ms. **(Chiappa 1997, p. 164)**

11. **(A):** Rarefaction tends to have a higher amplitude of wave I. Cochlear microphonics unlike wave I change polarity with changing click polarity. Alternating between rarefaction and condensation helps in getting rid of cochlear microphonics. **(Chiappa 1997, p. 171)**

12. **(C):** BAEPs would be normal in a patient with cortical deafness because the responses are recorded from the brainstem, not the cortex. Wave V is the last visible peak with progressive decrease of stimulus intensity. It is present but of a low amplitude than the contralateral ear recording. **(Chiappa 1997, p. 173)**

13. **(C):** Wave I appears after 1.4 ms. It is unmeasurable from the contralateral ear electrode. Cochlear microphonics, not wave I, changes polarity with changing click polarity. Alternating between rarefaction and condensation helps in getting rid of cochlear microphonics. **(Chiappa 1997, p. 171)**

14. **(B):** Wave III is usually present equidistant between wave I and V and it is the last wave to be present in that are with decreasing click intensities. Wave III is markedly attenuated in contralateral recordings. **(Chiappa 1997, p. 173)**

15. **(A):** Absolute latencies of all BAEP waves are increased with increasing click rate. Wave V is positive at vertex and ipsilateral earlobe sites. Wave V configuration reaches adult configuration by 3 to 6 months of age. **(Chiappa 1997, pp. 182, 189–192)**

16. **(C):** Infant BAEP sweep duration should be longer (15 to 20 ms) because of slower response seen in this age-group. **(Chiappa 1997, p. 269)**

17. **(C):** Wave I of BAEP is a negative wave recorded at the ear. Wave I is generated by distal 8th nerve action potentials. Wave III is produced in the superiorolivary complex in lower pons. Wave V is generated primarily in the lateral lemniscus (midbrain). **(Chiappa 1997, p. 200)**

References

1. Chiappa KH. *Evoked potentials in clinical medicine*, 3rd ed. Lippincott-Raven; 1997.

CHAPTER 10

Somatosensory Evoked Potentials

QUESTIONS

1. Somatosensory evoked potentials are best recorded by using an amplification of:
   A. 100
   B. 1,000
   C. 10,000
   D. 100,000

2. Somatosensory evoked potentials are best recorded using a low-frequency filter of:
   A. 30 Hz
   B. 3 Hz
   C. 50 Hz
   D. 5 Hz

3. Somatosensory evoked potentials are best recorded by using a high-frequency filter of:
   A. 100 Hz
   B. 1,000 Hz
   C. 3,000 Hz
   D. 30,000 Hz

4. Somatosensory evoked potentials are best recorded by using a sweep duration of:
   A. 10 ms
   B. 20 ms
   C. 200 ms
   D. 100 ms

5. **Erb's point N9 is the reference point for latency measurement. It corresponds to:**
   A. Cervicomedullary junction
   B. Cauda equina
   C. Brachial plexus
   D. Cubital fossa

6. **N11-13 complex corresponds to:**
   A. Cervicomedullary junction
   B. Cauda equine
   C. Brachial plexus
   D. Cubital fossa

7. **N20 shown in the Figure 10.1 corresponds to:**
   A. Cervicomedullary junction
   B. Cauda equine
   C. Thalamus
   D. Cubital fossa

8. **Somatosensory evoked potential (SSEP) in lower extremity is commonly done using posterior tibial nerve compared to common peroneal nerve at the knee because:**
   A. Posterior tibial nerve stimulation has high threshold for stimulation
   B. Posterior tibial nerve stimulation produce higher amplitude lumbar potentials
   C. Posterior tibial nerve is a pure sensory nerve
   D. All of the above

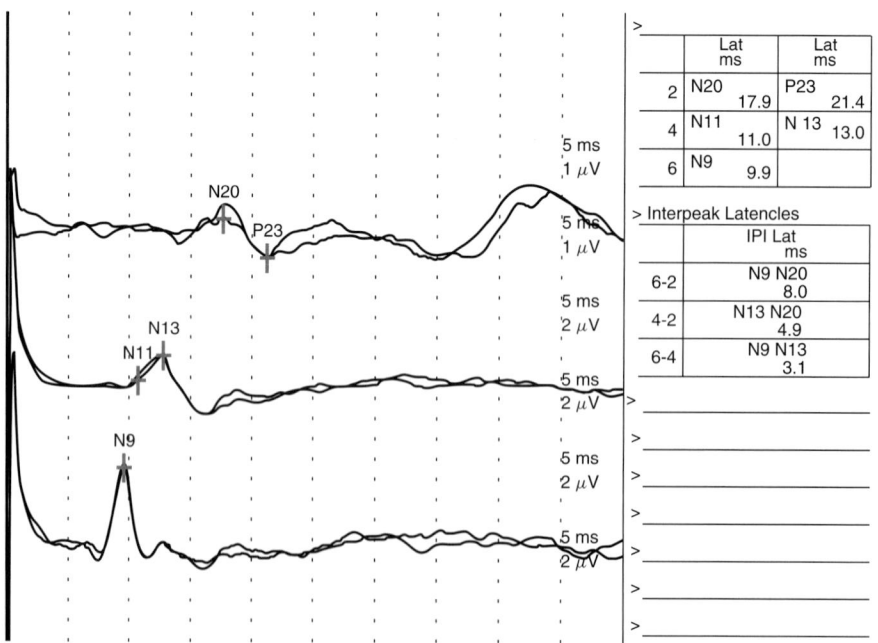

Figure 10.1. Somatosensory evoked potential (SSEP) median

9. SSEP tests the integrity of:
   A. Corticospinal tract
   B. Spinothalamic tract
   C. Spinocerebellar tract
   D. Dorsal column

10. The most common problem encountered with the SSEP test is excessive muscle activity.
    A. False
    B. True

11. The SSEP stimulus intensity should be enough to produce muscle twitch.
    A. True
    B. False

12. If the SSEP stimulus intensity is low, the scalp potentials will be affected as follows:
    A. Responses will have a low amplitude
    B. Responses will have a high amplitude
    C. Responses will have a short latency
    D. Responses will have a prolonged latency

13. In SSEP, bilateral stimulation is not used in clinical practice because:
    A. Amplitude of responses will be low
    B. Responses will be delayed
    C. Normal responses from the good side will mask the abnormality on the bad side
    D. None of the above

14. Bilateral stimulation is used in the following situation:
    A. To monitor spinal cord functioning during surgery
    B. To evaluate brain death
    C. To evaluate multiple sclerosis
    D. A and B

15. SSEP waveform amplitudes and signal-to-noise ratio are quite low. It is necessary to use at least:
    A. 100 repetitions
    B. 1,000 repetitions
    C. 10,000 repetitions
    D. 100,000 repetitions

16. For better resolution of unclear SSEP waveforms, all procedures are used except:
    A. Increasing stimulus repetition to 2,000
    B. Performing multiple trials
    C. Decreasing stimulus intensity
    D. Sedation of the patient

17. **The following waveform identification in SSEP is correct except:**
    A. Brachial plexus potential appears as a diphasic positive–negative waveform N9
    B. Cervicomedullary potential appears as a negative–positive deflection N11/13
    C. Thalamocortical activity produced by lower limb stimulation appears as a negative waveform P37
    D. Lumbar activity in the cauda equina appears as a positive–negative deflection N21

18. **Patient's height must be recorded for:**
    A. Visual evoked potential (VEP)
    B. Median SSEP
    C. Ulnar SSEP
    D. Posterior tibial SSEP

19. **Spinal cord values with SSEP in children reach the adult values by age:**
    A. 1 year
    B. 3 years
    C. 5 years
    D. 10 years

20. **Abnormal interpeak latency EP-N13 indicates:**
    A. Conduction defect in the large-fiber sensory system, central to the brachial plexus both below and above medulla (but caudal to thalamus)
    B. Conduction defect in the large-fiber sensory system above the lower medulla
    C. Conduction defect in the large-fiber sensory system, central to the brachial plexus and at or below the lower medulla
    D. Conduction defect in the large-fiber sensory system, central to brachial plexus and below the lower medulla

21. **Absence or very low amplitude of N13 indicates:**
    A. Conduction defect in the large-fiber sensory system, central to the brachial plexus both below and above medulla (but caudal to thalamus)
    B. Conduction defect in the large-fiber sensory system above the lower medulla
    C. Conduction defect in the large-fiber sensory system above the lower medulla and below thalamus.
    D. Conduction defect in the large-fiber sensory system, central to the brachial plexus and at or below the lower medulla

22. **N13 is normal but N20 is absent, this indicates:**
    A. Conduction defect in the large-fiber sensory system, central to the brachial plexus both below and above medulla (but caudal to thalamus)
    B. Conduction defect in the large-fiber sensory system above the lower medulla
    C. Conduction defect in the large-fiber sensory system above the lower medulla and below thalamus.
    D. Conduction defect in the large-fiber sensory system, central to the brachial plexus and at or below the lower medulla

23. **Abnormal interpeak latency of EP-N13 and N13-N20 indicates:**
    A. Conduction defect in the large-fiber sensory system, central to the brachial plexus, both below and above the medulla (but caudal to thalamus)
    B. Conduction defect in the large-fiber sensory system above the lower medulla
    C. Conduction defect in the large-fiber sensory system above the lower medulla and below thalamus
    D. Conduction defect in the large-fiber sensory system, central to the brachial plexus and at or below the lower medulla

24. **In newborns, SSEP responses are best recorded with the following except:**
    A. Erb's potential is best recorded 1 cm above axilla
    B. Bilateral nerve stimulation is not necessary
    C. Sweep duration should be 80 to 100 ms
    D. Lower stimulus rate (1.1 Hz) produces higher-amplitude SSEPs

25. **Features of far field potentials include all except:**
    A. Low amplitude
    B. Long latency
    C. High frequency
    D. Wide distribution over the scalp

26. **In the recording shown in Figure 10.2, which of the following is the most reliable cortical potential?**
    A. N45
    B. P37
    C. N21
    D. Popliteal fossa (PF)

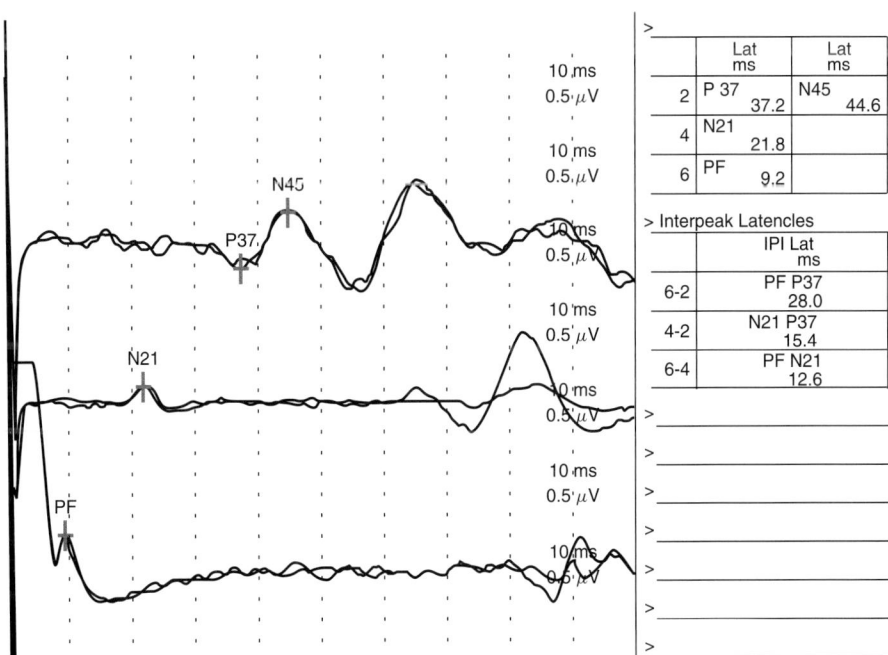

Figure 10.2. Somatosensory evoked potential (SSEP) tibial

ANSWERS

1. (D): Averaging in evoked potential testing increases the signal-to-noise ratio. An amplification of 100,000 is needed to improve the signal-to-noise ratio. **(Chiappa 1997, p. 288)**

2. (A): Low-frequency filter of 30 Hz is appropriate to record SSEP. This frequency cutoff will filter movement artifact encountered during the recordings. **(Chiappa 1997, p. 288)**

3. (C): A high-frequency filter of 3,000 Hz is most appropriate for recording SSEP. **(Chiappa 1997, p. 288)**

4. (D): Sweep duration of 100 ms is needed to record SSEP. Also, the dwell time should be set at 0.2 ms. **(Chiappa 1997, p. 288)**

5. (C): The N9 of the somatosensory evoked potential originates in the brachial plexus. Cauda equina potential corresponds to N21. N11/13 complex corresponds to cervicomedullary junction. Cubital fosse potentials are not used in SSEP in clinical practice. **(Chiappa 1997, p. 288)**

6. (A): The N9 of the somatosensory evoked potential originates in the brachial plexus. Cauda equina potential corresponds to N21. N11/13 complex corresponds to cervicomedullary junction. Cubital fosse potentials are not used in SSEP in clinical practice. **(Chiappa 1997, pp. 289–290)**

7. (C): The N9 of the somatosensory evoked potential originates in the brachial plexus. Cauda equina potential corresponds to N21. N11/13 complex corresponds to cervicomedullary junction. Cubital fosse potentials are not used in SSEP in clinical practice. N20 corresponds to thalamic action potential. **(Chiappa 1997, pp. 289–290)**

8. (B): Stimulation of posterior tibial nerve at ankle produces higher amplitude of lumbar and scalp potentials than stimulation of common peroneal nerve. This amplitude difference is possibly due to rich innervation of the sole of the foot. **(Chiappa 1997, p. 294)**

9. (D): SSEP only tests the integrity of myelinated large fibers sensory system, that is, dorsal column. At times, SSEP may be completely normal if the lesion involves the anterior two third of the spinal cord and spares dorsal column. **(Chiappa 1997, p. 319)**

10. (B): Muscle artifact is the most common problem encountered. It is very crucial that patient relaxes during the testing. If the problem persists, a short acting hypnotic can be given to improve signal-to-noise ratio. **(Chiappa 1997, p. 303)**

11. (A): SSEP stimulus intensity should be enough to produce a muscle twitch. Inadequate stimulation of the tibial nerve sometimes results in delayed responses. **(Chiappa 1997, p. 303)**

12. (D): If stimulus intensity is increased, the responses' latency will become normal. SSEP stimulus intensity should be enough to produce a muscle twitch. Inadequate stimulation of the tibial nerve sometimes results in delayed responses. **(Chiappa 1997, p. 306)**

13. (C): SSEP bilateral stimulation is not used in clinical practice because of normal responses from good side will mask the abnormality on the bad side. The only possible use for bilateral stimulation in SSEP testing may be for obtaining a larger amplitude response when using the lower limb SSEP to monitor spinal cord functioning during surgery. **(Chiappa 1997, p. 306)**

14. (C): Bilateral absence of cortical responses (N20) on median somatosensory evoked potential is associated with a poor outcome in comatose patients. The studies suggest that if a comatose patient is missing cortical responses bilaterally, then that patient's outcome will be a persistent vegetative state at best. **(Chiappa 1997, p. 389)**

15. (B): It is necessary to use 1,000 or more stimulus repetitions to ensure waveform clarity. **(Chiappa 1997, p. 306)**

16. (C): Increasing stimulus repetition to 2,000 is very helpful. Patient may be sedated to improve the signal-noise-ratio. Performing multiple trials will improve the signal-noise-ratio. Decreasing the stimulus intensity will decrease the signal-noise-ratio. **(Chiappa 1997, p. 308)**

17. (C): The N9 of the somatosensory evoked potential originates in the brachial plexus, appears as positive–negative waveform. Cauda equina potential corresponds to a positive–negative deflection as N21. N11-13 complex corresponds to cervicomedullary junction appearing as negative–positive waveform. P37 appears as positive waveform (not as negative) and corresponds to thalamocortical activity produced by lower limb stimulation. **(Chiappa 1997, p. 309)**

18. (D): Patient's height must be recorded because it is used to correct the lower limb absolute latencies or lumbar to cerebral conduction times for distance traveled. In all other options, height is not needed. **(Chiappa 1997, p. 310)**

19. (C): Newborns have both spinal cord and peripheral nerve conduction velocities delayed and these velocities progressively increase with age. The peripheral nerve conduction velocity reaches the adult value by 3 years. However, the spinal cord central conduction time reaches adult value by 5 years. **(Chiappa 1997, p. 336)**

20. (D): Abnormal interpeak latency EP-N13 indicates conduction defect in the large-fiber sensory system, central to brachial plexus and below the lower medulla. **(Chiappa 1997, p. 363)**

21. (D): Very low amplitude of N13 indicates a conduction defect in the large-fiber sensory system, central to the brachial plexus and at or below the lower medulla. **(Chiappa 1997, p. 363)**

22. (B): An absent N20 in the presence of normal N13 indicates a conduction defect in the large-fiber sensory system above the lower medulla. **(Chiappa 1997, p. 364)**

23. (A): Abnormal interpeak latencies of Ep-N13 and N13-N20 indicate a conduction defect in the large-fiber sensory system, central to the brachial plexus both below and above the medulla (but caudal to thalamus). **(Chiappa 1997, p. 365)**

24. (B): In premature and newborn infants, bilateral nerve stimulation is necessary to obtain reproducible results. **(Chiappa 1997, p. 456)**

25. (B): Far-field potentials actually have short latencies. They have a wide distribution, high frequency, and low amplitude. **(Chiappa 1997, p. 604)**

26. (B): P37 and N45 are both cortical potentials recorded with tibial nerve stimulation. P37 potential is commonly used as the more reliable cortical potential because of easier recognition and better reproducibility. Also, most normative data are based on P37 measurements. **(Chiappa 1997, pp. 317–318)**

References

1. Chiappa KH. *Evoked potentials in clinical medicine*, 3rd ed. Lippincott-Raven; 1997.

CHAPTER 11

Evoked Potentials in Clinical Practice

QUESTIONS

1. **Abnormal P100 latency can normalize in rare cases.**
   A. True
   B. False

2. **The findings seen in Figure 11.1 show a delay in P37 latency. This is indicative of a:**
   A. Conduction defect in the large-fiber sensory system above cauda equina and below the sensory cortex
   B. Conduction defect in the large-fiber sensory system, peripheral to the cauda equina
   C. Conduction defect in the large-fiber sensory system, peripheral to cauda equina and below the sensory cortex
   D. All of the above

3. **The abnormal I-III interpeak latency is commonly seen with:**
   A. Multiple sclerosis
   B. Acoustic neuroma
   C. Adrenoleukodystrophy
   D. Transverse myelitis

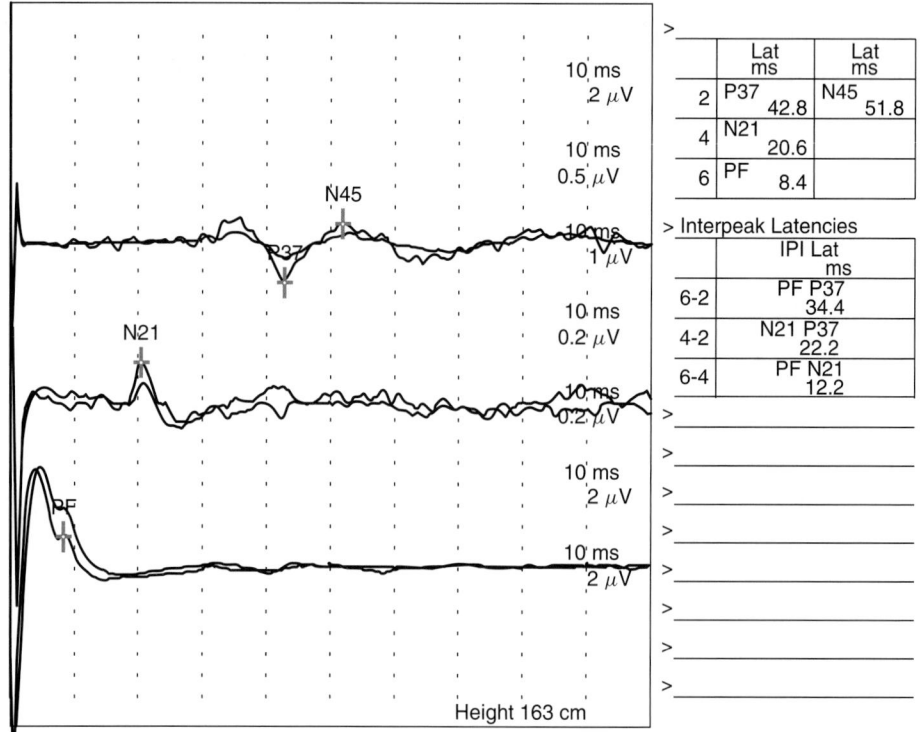

Figure 11.1. Somatosensory evoked potential (SSEP) delay

4. Bilaterally delayed P100 latency can localize the lesion only anterior to the optic chiasm.
   A. True
   B. False
   C. Bilateral delayed P100 is never seen clinically
   D. Bilateral delayed P100 is pathognomonic for multiple sclerosis

5. A 29-year-old patient presents with new onset of paraplegia. Magnetic resonance imaging (MRI) of the spine shows T2 high-intensity signal in the cord, but brain MRI is negative. Further work-up should include all of the following tests except:
   A. Evoked potential
   B. Echocardiogram
   C. Lumbar puncture (LP)
   D. Nerve biopsy

6. Brain death is associated with which of the following brainstem auditory evoked potential (BAEP) abnormalities?
   A. Absence of bilateral wave I
   B. Absence of bilateral wave III

C. Absence of all waveforms following wave I bilaterally
D. All of the above

7. **Delayed-lumbar potential N21 but normal N21-P37 interpeak latency indicates:**
   A. Conduction defect in the large-fiber sensory system above cauda equina and below the sensory cortex
   B. Conduction defect in the large-fiber sensory system, peripheral to the cauda equina
   C. Conduction defect in the large-fiber sensory system, peripheral to cauda equina and below the sensory cortex
   D. All of the above

8. **The tracing shown in Figure 11.2 is that of visual evoked potential (VEP) with right and left eye stimulation. These findings localize the lesion:**
   A. Anterior to the optic chiasm
   B. Posterior to the optic chiasm
   C. Cannot localize the lesion
   D. Is a normal study

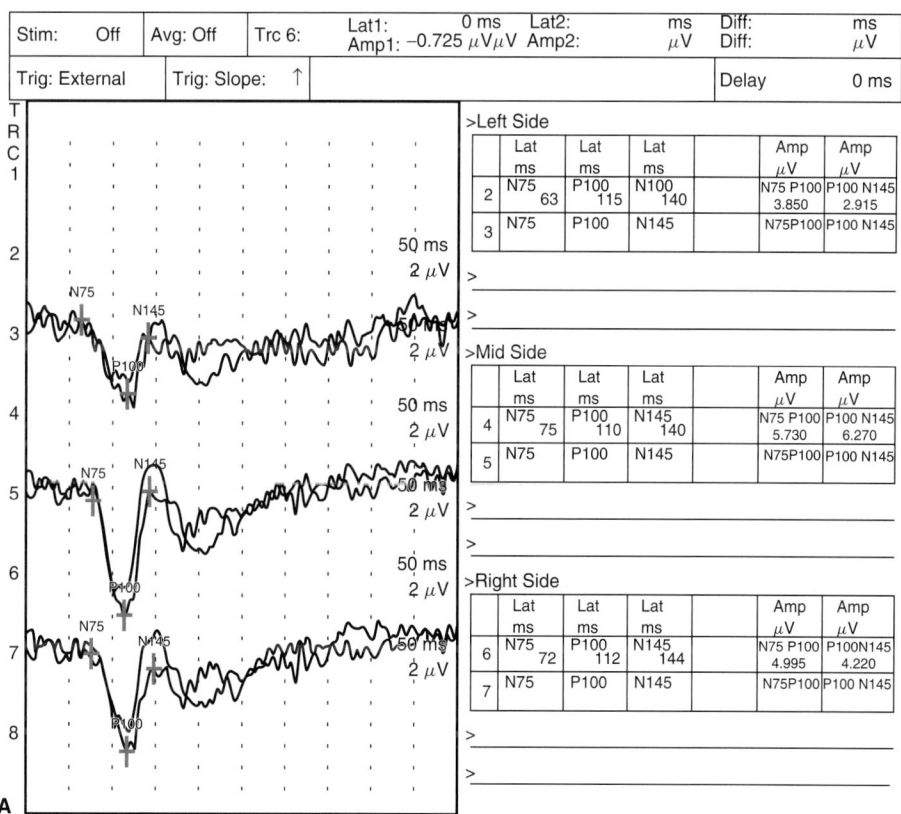

Figure 11.2. Visual evoked potential (VEP)-right (**A**) and VEP-left (**B**)

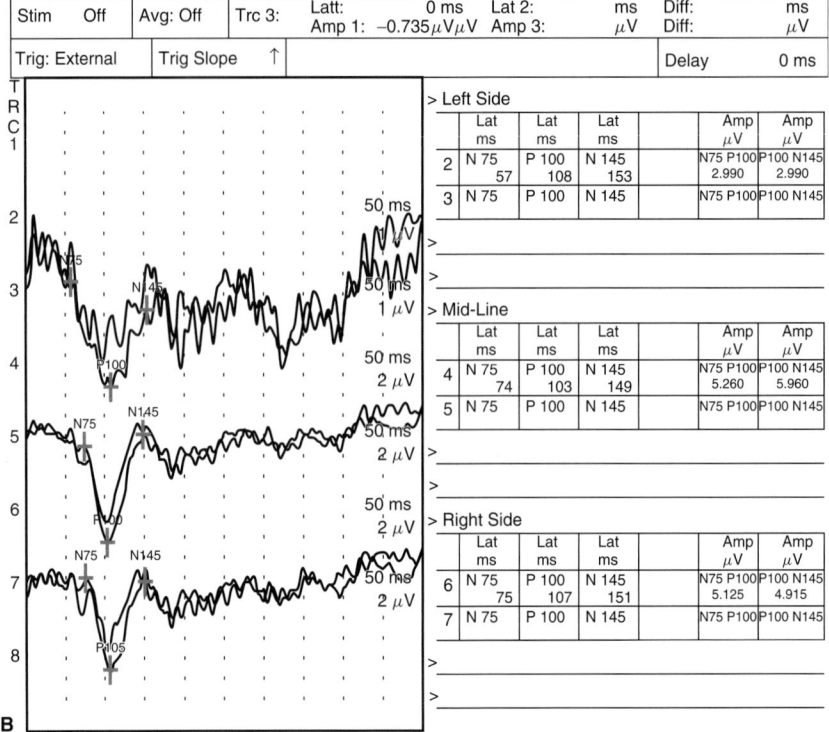

Figure 11.2. (Continued)

9. **Abnormal interpeak latency of N21-P37 indicates:**
   A. Conduction defect in the large-fiber sensory system above cauda equina and below the sensory cortex
   B. Conduction defect in the large-fiber sensory system, peripheral to the cauda equina
   C. Conduction defect in the large-fiber sensory system, peripheral to cauda equina and below the sensory cortex
   D. All of the above

10. **Absence of wave V suggests:**
    A. Conduction defect between distal eighth nerve and lower pons
    B. Conduction defect between lower pons and medulla
    C. Conduction defect rostral to lower pons
    D. Conduction defect between cervicomedullary junction and midbrain

11. **If lumbar potential N21 is absent but absolute latency of P37 is normal, then the test cannot be interpreted.**
    A. True
    B. False

12. **Abnormal interpeak latency N13-N20 indicates:**
    A. Conduction defect in the large-fiber sensory system, central to the brachial plexus both below and above medulla (but caudal to thalamus)
    B. Conduction defect in the large-fiber sensory system above the lower medulla
    C. Conduction defect in the large-fiber sensory system above the lower medulla and below the thalamus

D. Conduction defect in the large-fiber sensory system, central to the brachial plexus and at or below the lower medulla

13. **In barbiturate coma, when electroencephalographic (EEG) finding is isoelectric, the somatosensory evoked potentials (SSEPs) show little change.**
    A. True
    B. False

14. **The finding shown in Figure 11.3 at P100 is:**
    A. Always an artifact
    B. Often a normal variant
    C. Diagnostic of blindness
    D. Usually considered abnormal

15. **If brachial plexus potential cannot be recorded, but absolute latencies of N13 and N20 are normal, then the test can be interpreted as normal.**
    A. True
    B. False

16. **The finding shown in Figure 11.4:**
    A. Localizes the lesion anterior to the optic chiasm
    B. Cannot localize the lesion
    C. Localizes in inferior colliculus
    D. Localizes only to occipital cortex

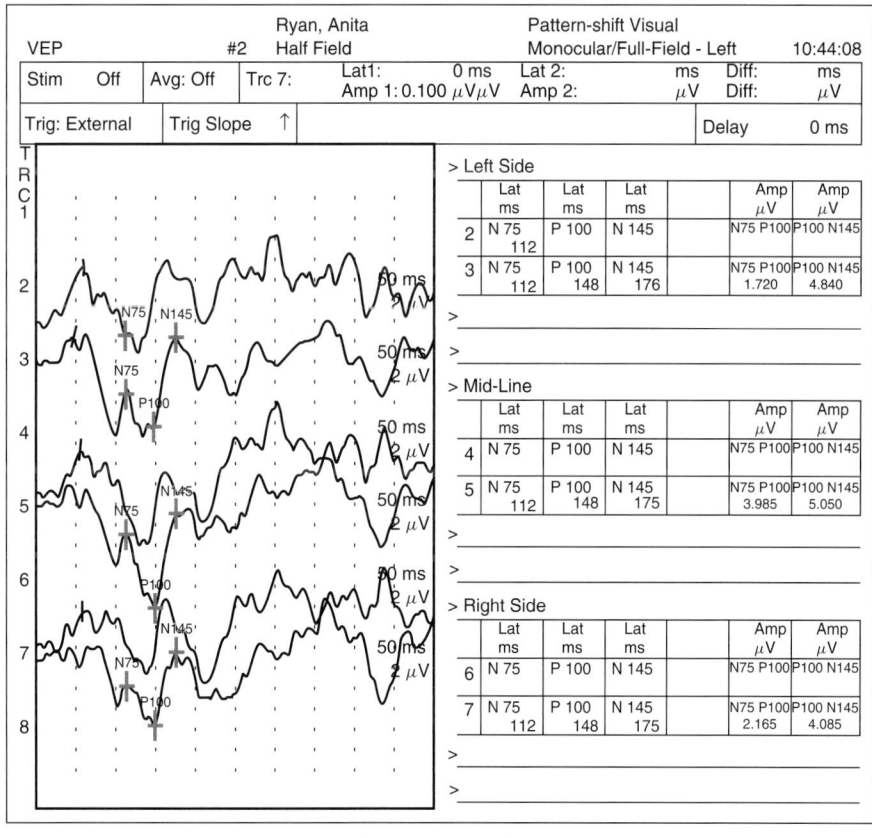

Figure 11.3. Visual evoked potential (VEP)-bifid

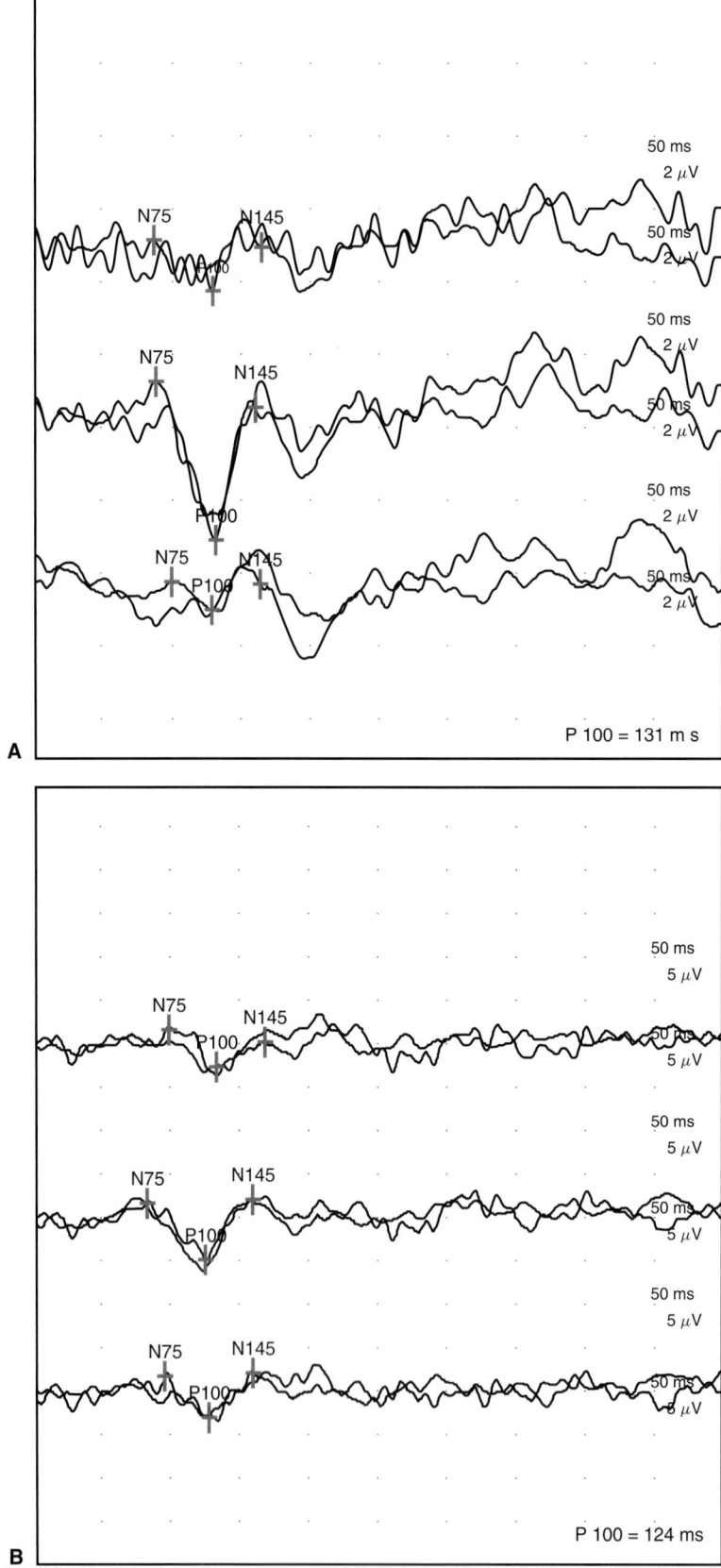

Figure 11.4. Visual evoked potential right side (A) and left side (B)

17. **Delayed-lumbar potential N21 and N21-P37 interpeak latency indicate:**
    A. Conduction defect in the large-fiber sensory system above cauda equina and below the sensory cortex
    B. Conduction defect in the large-fiber sensory system, peripheral to the cauda equina
    C. Conduction defect in the large-fiber sensory system, peripheral to cauda equina and below the sensory cortex
    D. All of the above

ANSWERS

1. **(A):** There have been some rare reports of P100 latency normalizing with or without clinical recovery. Though a rare phenomenon, it has been seen more commonly in children. (**Kriss, A et al. J Neurol Neurosurg Psych 1988; 51, pp. 1253–1258**)

2. **(A):** The delayed P37 absolute latency and abnormal interpeak latency of N21-P37 indicate conduction defect in the large-fiber sensory system above cauda equina and below the sensory cortex. (**Chiappa 1997, p. 366**)

3. **(B):** The abnormal I-III interpeak latency is commonly seen with acoustic neuroma, while multiple sclerosis and adrenoleukodystrophy can have abnormal III-V latency. BAEP is usually normal in transverse myelitis. (**Chiappa 1997, p. 210**)

4. **(B):** Bilateral delayed P100 cannot be localized because it can be seen with various pathologies, for example, retinal degenerations, optic chiasm tumors, central nervous system (CNS) degenerative diseases, and bilateral optic radiation lesions. (**Chiappa 1997, p. 99**)

5. **(D):** Nerve biopsy will not be helpful in the case of a myelopathy. However, evoked potential may help in finding other silent lesions in visual or auditory pathways suggestive of multiple sclerosis. LP will help in evaluation of inflammatory or autoimmune myelopathy, while electroencephalography (ECG) will help in ischemic myelopathy. (**Chiappa 1997, p. 368**)

6. **(C):** Without wave I present bilaterally, no inference can be made. However, in the presence of wave I, absence of BAEP waves V is compatible with brain death. (**Chiappa 1997, p. 238**)

7. **(B):** Delayed-lumbar potential N21, but normal N21-P37 interpeak latency indicates conduction defect in the large-fiber sensory system, peripheral to the cauda equine. (**Chiappa 1997, p. 366**)

8. **(A):** This tracing shows normal P100 absolute latency but the interside difference is significant. Unilateral P100 delay is seen with prechiasmal lesion. A unilateral P100 abnormality indicates an ipsilateral lesion of the visual pathway anterior to the optic chiasm such as a unilateral demyelinating process or optic nerve glioma. (**Chiappa 1997, p. 99**)

9. (A): Abnormal interpeak latency of N21-P37 indicates conduction defect in the large-fiber sensory system above cauda equina and below the sensory cortex. **(Chiappa 1997, p. 366)**

10. (C): Absence of wave V is indicative of conduction defect in the brain stem auditory system rostral to the pons. **(Chiappa 1997, p. 211)**

11. (B): The loss of lumbar potential N21 can be due to technical reasons. If the absolute latency of P37 is normal then the test can be interpreted as normal. **(Chiappa 1997, p. 366)**

12. (C): Abnormal interpeak latency indicates a conduction defect in the large-fiber sensory system above the lower medulla and below thalamus. **(Chiappa 1997, p. 364)**

13. (A): If barbiturates are given in doses sufficient to render the EEG findings isoelectric, little change is seen in the SSEP because it is very resistant. Cortical response amplitude in SSEP may decrease to some extent but is not isoelectric, while EEG is very sensitive to barbiturates. **(Chiappa 1997, p. 337)**

14. (D): This tracing shows a bifid P100 waveform. Bifid P100 is very rarely seen in healthy individuals. It may be produced with upper visual field contributing inverted-polarity activity to the inion resulting in bifid waveform. Using only the lower field stimulation may correct the waveform. **(Chiappa 1997, p. 102)**

15. (A): The loss of brachial plexus potential can be due to technical reasons. If the absolute latencies of N13 and N20 are normal then the test can be interpreted as normal. **(Chiappa 1997, p. 365)**

16. (B): This tracing shows bilaterally delayed P100 latency. Bilaterally delayed P100 cannot be localized as it can be seen with various pathologies, for example, retinal degenerations, optic chiasm tumors, CNS degenerative diseases, and bilateral optic radiation lesions. Hence with bilateral P100 delayed, it is mentioned that due to the bilateral delay it cannot be further localized. **(Chiappa 1997, p. 99)**

17. (C): Delayed-lumbar potential (N21) and N21-P37 interpeak latency indicates conduction defect in the large-fiber sensory system peripheral to cauda equina and below the sensory cortex. **(Chiappa 1997, p. 366)**

References

1. Chiappa KH. *Evoked potentials in clinical medicine*, 3rd ed. Lippincott-Raven; 1997.

SECTION 4

Nerve Conduction Studies

CHAPTER 12

General Principles of Nerve Conduction Studies

QUESTIONS

1. **If a routine sural nerve sensory study shows an absent potential while the patient has a normal sensory examination of the lateral foot, one should consider:**
   A. Tarsal tunnel syndrome
   B. Amyotrophic lateral sclerosis (ALS)
   C. Technical problem
   D. Absent sural nerve

2. **All statements about dorsal root ganglion (DRG) are correct except:**
   A. It contains the cell bodies of 1 C fibers
   B. It consists of bipolar cells with two separate axonal projections
   C. The projections of DRG enter the spinal cord on the dorsal side
   D. Peripheral projections of the DRG become the sensory fibers in the peripheral nerve

3. **The following statements about myelination are correct except:**
   A. Myelin is present on all sensory conducting fibers
   B. Myelin is derived from Schwann cells, the major supporting cells in the peripheral nervous system
   C. For every myelinated fiber, successive segments are myelinated by single Schwann cell
   D. The axonal member is exposed near the neuromuscular junction

4. **All statements are correct except:**
   A. Amplitude is usually measured from baseline to the negative peak
   B. Controlled motor action potential (CMAP) duration is a measure of synchrony (simultaneous firing of individual muscle fibers)
   C. Conduction velocity is a measure of the speed of the fastest conducting motor axons
   D. When the nerve is stimulated proximally, the CMAP amplitude is approximately half the amplitude of the distal stimulation waveform

5. **Cool temperatures of the limb result in a change in nerve conduction study as:**
   A. Increase in conduction velocity
   B. Decrease in conduction velocity
   C. Decrease in F-wave latency
   D. Decrease in amplitude

6. **The following maneuvers will help decrease the stimulus artifact except:**
   A. Ground electrode should not be placed between the stimulator and the recording electrodes
   B. Reducing electrodes impedance mismatch
   C. Rotate the anode of the stimulator while maintaining the position of the cathode
   D. Stimulator and recording electrode cables do not overlap

7. **The rate of axonal regrowth is:**
   A. Similar to the rate of demyelinating regrowth
   B. Similar for distal upper and lower extremities
   C. Is approximately 1 cm per day
   D. Is approximately 1 mm per day

8. **An acute axonal injury is defined as:**
   A. Injury within few hours
   B. Injury within few days
   C. Injury within few weeks
   D. Injury within few months

9. **An absent sural nerve sensory study with intact sensation over the lateral foot most likely represents:**
   A. An chronic axonal lesion to the sural nerve
   B. A chronic demyelinating lesion to the sural nerve
   C. Collateral innervation of the lateral foot
   D. A technical problem

10. **All statements regarding the response in Figure 12.1 are incorrect except:**
    A. Recorded with routine sensory nerve stimulation studies
    B. Recorded with routine motor nerve stimulation studies
    C. Attenuated in a cold limb
    D. Attenuated in a warm limb

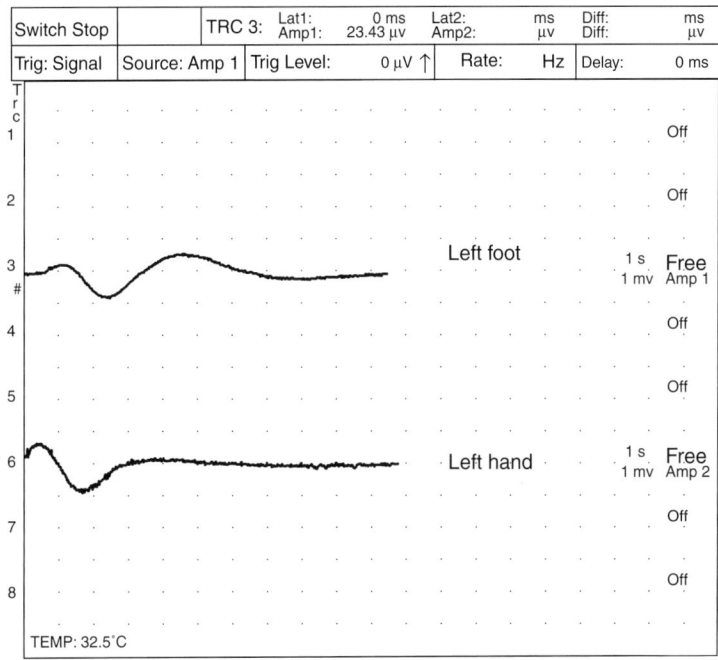

Figure 12.1. Sympathetic skin response

11. **Nerve conduction studies in myopathic disorder are:**
    A. Sensory conduction studies are always normal
    B. Motor nerve conduction studies are always normal
    C. F-wave is always normal
    D. Repetitive nerve stimulation is normal

ANSWERS

1. (C): This is most commonly a technical problem, for example, amplifier turned off, electrode placement, skin impedance, previous sural nerve biopsy, edema, or ankle deformity. One should check the integrity of the circuit before attributing the abnormality to a disease. Amyotrophic lateral sclerosis (ALS) is a motor neuron disease and sensory nerves are usually preserved. Tarsal tunnel syndrome (planter nerve entrapment) would not affect the sural sensory response. **(Preston and Shapiro 1998, p. 6)**

2. (A): Dorsal root ganglion contains the cell bodies of large myelinated sensory 1A fibers. It consists of bipolar cells with two separate axonal projections. The projections of dorsal root ganglion enter the spinal cord at the entry zone of the dorsal funiculus. Peripheral projections of the DRG eventually become the sensory fibers in the peripheral nerve. **(Preston and Shapiro 1998, p. 9)**

3. (A): Only large fast conducting fibers are myelinated. Myelination of the peripheral nerves is the function of the Schwann cells and every segment is myelinated by a single Schwann cell. The myelination is absent near the neuromuscular junction and at nodes of Ranvier. **(Preston and Shapiro 1998, p. 12)**

4. (D): When the nerve is stimulated proximally the CMAP area, amplitude, and duration are similar to those of the distal stimulation waveform. CMAP amplitude is usually measured from baseline to the negative peak and less commonly from the first negative peak to the next positive peak. CMAP duration is a measure of synchrony (simultaneous firing of individual muscle fibers). Conduction velocity is a measure of the speed of the fastest conducting motor axons, which is calculated by dividing the distance traveled by the nerve conduction time from a proximal to a distal stimulation site. One cannot calculate a true distal motor conduction velocity by using a distal motor latency and a distance; as the velocity measurement would include the conduction time through the neuromuscular junction and along the muscle fiber. (**Preston and Shapiro 1998, p. 26**)

5. (B): Cool temperatures result in slowing of conduction velocity, which is prominent in large rather than small myelinated fibers. For motor and sensory studies, conduction velocity slows between 1.5 and 2.5 m per second for every 1°C drop in temperature and distal latency prolongs by approximately 0.2 ms per degree. Cooling of the limb results in a larger amplitude and longer duration potential, associated with a longer distal latency and slower conduction velocity. (**Preston and Shapiro 1998, p. 85**)

6. (A): Placing ground electrode between the stimulator and the recording electrodes helps reduce the stimulus artifact. Reducing electrodes impedance will decrease the stimulus. Rotating the anode of the stimulator while maintaining the position of the cathode will decrease the stimulus. Keeping the stimulator and recording electrode cables separate and kept as far as possible will help reduce the influence of stimulus artifact. (**Preston and Shapiro 1998, p. 90**)

7. (D): The rate of axonal regrowth is limited by slow axonal transport. It is approximately 1 mm per day, so an upper extremity axonal lesions will have a faster rate of recovery compared to lower extremities. It is much slower than the remyelination rate, which only takes few weeks. Therefore, a demyelinating lesion with secondary axonal loss will have a more guarded prognosis than a purely demyelinating lesion. (**Kimura Electrodiagnosis in Diseases of Nerve and Muscle: Principles and Practice 1989**)

8. (C): Hyperacute injury is defined as injury within 1 week, acute injury is within few weeks, subacute injury is within weeks to few months, and chronic injury is defined as injury for more than few months. Enough time is passed for wallerian degeneration to have occurred. Accordingly, nerve conduction studies are abnormal, amplitudes are decreased with relatively normal conduction velocities and distal latencies, unless some of the largest and fastest axons have been lost, in which case some slowing of velocity and latency occurs. (**Preston and Shapiro 1998, p. 213**)

9. (D): Technical problems can easily lead to abnormal or absent responses. The key in determining the difference between a technical abnormality and a true lesion is the lack of clinical electrophysiologic correlation. The study is best done with patient lying on his side. Side-to-side comparison of amplitude and latency are often helpful. The examiner should investigate possible technical causes and should retest the nerve several times before declaring the result abnormal. (**Preston and Shapiro 1998, p. 134**)

10. (C): The images in Figure 12.1 depict a normal sympathetic skin response from left hand and foot. Sympathetic skin response is sensitive to temperature and may be attenuated in a cold limb. Warming the limb helps in getting a better resolution of the sympathetic skin response. Sympathetic skin response cannot be obtained from routine motor or sensory studies because it requires a separate set up. (**Doston 1997, pp. 32–45**)

11. (B): Routine nerve conduction studies should always be done in patients with suspected myopathy. Sensory conduction studies are always normal unless there is a coexistent neuropathy. Routine motor conduction studies are usually normal, as myopathies usually involve nerves supplying the proximal muscles that are not tested in routine motor nerve conduction studies. However, if the myopathy is severe enough to affect distal and proximal muscles or one of the rare myopathies that preferentially affects distal muscles, motor nerve. (**Preston and Shapiro 1998, p. 527**)

References

1. Dotson R. *J Clin Neurophysiol*. 1997;14(1):32–45.
2. Preston DC, Shapiro BE. *Electromyography and Neuromuscular Disorders*, 1st ed. Butterworth-Heineman; 1998.

CHAPTER 13

Sensory Nerve Conduction Studies

QUESTIONS

1. **For sensory conduction studies, the gain is usually set at:**
   A. 1 μV
   B. 20 μV
   C. 100 μV
   D. 200 μV

2. **The following statements are correct except:**
   A. The recording electrodes are placed in line over the nerve with an interelectrode distance of 3 to 4 cm
   B. Most sensory nerves require a current in the range of 5 to 30 mA to achieve supramaximal stimulation
   C. Sensory nerves require a higher threshold for stimulation than do motor fibers
   D. In sensory studies, a conduction velocity can be calculated using one stimulation site alone

3. **Antidromic recording of sensory nerves is superior to orthodromic technique because:**
   A. Amplitude is higher using antidromic stimulation
   B. The electrodes are closer to the nerve
   C. It is less subject to noise or other artifact
   D. The entire nerve is stimulated including the motor fibers

4. **Myelinated fibers conduct at a velocity of approximately:**
   A. 100 m/s
   B. 65 m/s

C. 30 m/s
D. 10 m/s

5. **Demyelination is associated with:**
   A. Marked slowing of conduction velocity (slower than 85% of the lower limit of normal)
   B. Marked prolongation of distal latency (longer than 110% of upper limit of normal)
   C. Low compound motor action potential (CMAP) amplitude
   D. Drop in amplitude in proximal stimulation compared to the distal stimulation

6. **Routine nerve conduction studies are normal in ___% of patients with clinical symptoms and signs of carpal tunnel syndrome.**
   A. 5%
   B. 20%
   C. 30%
   D. 40%

7. **Carpal tunnel syndrome electrodiagnostic studies usually show (see Fig. 13.1):**
   A. Higher amplitude of median CMAP compared to ulnar nerve
   B. The median nerve orthodromic absolute onset sensory latency is delayed by >3.7 ms
   C. The median nerve orthodromic sensory latency is delayed by >2.0 ms compared to the ulnar nerve sensory distal latency
   D. The median nerve orthodromic sensory latency is delayed by >0.2 ms compared to the ulnar nerve sensory distal latency

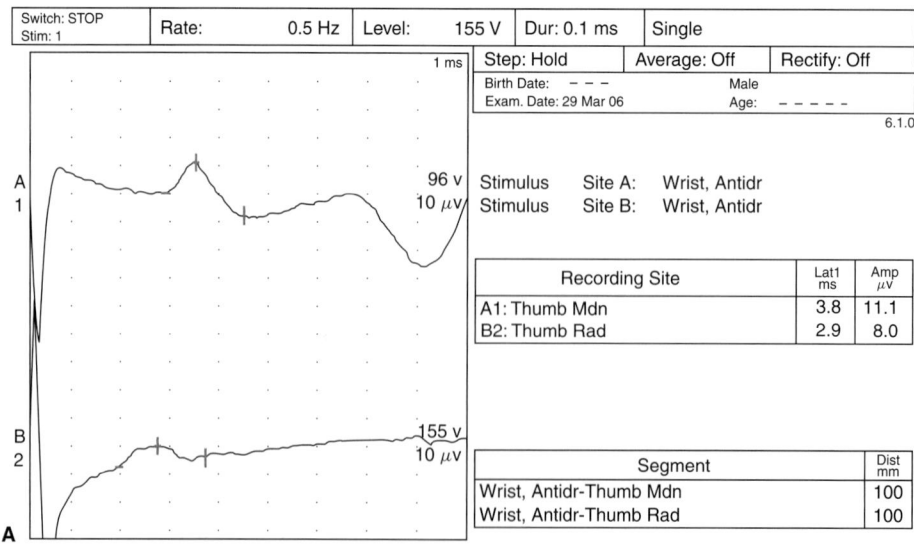

Figure 13.1. (**A**) Median versus radial sensory nerve conduction (SNC) and (**B**) median versus ulnar motor nerve conduction (MNC)

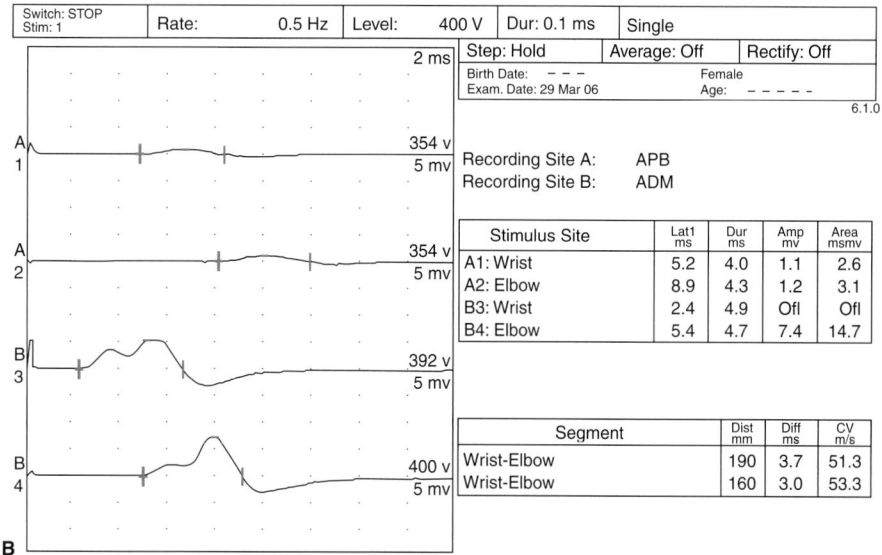

Figure 13.1. (Continued)

8. Median- versus ulnar-comparison tests are good tests to confirm carpal tunnel syndrome for the following reasons except:
   A. They create an ideal internal control
   B. Distance is different
   C. Temperature is constant
   D. Nerve fiber size is constant

9. Median- versus ulnar-sensory nerve short latency midpalmar study comparison test is abnormal if the median latency is delayed by more than ____ the ulnar latency.
   A. 0.1 ms
   B. 0.2 ms
   C. 0.3 ms
   D. 0.4 ms

10. In a patient with right foot pain, a sural and plantar sensory response is unmeasurable. The lesion can be present in the following except:
    A. Tarsal tunnel syndrome
    B. Sciatic lesion
    C. Lumbosacral plexus lesion
    D. Peripheral neuropathy

11. All of the following statements are correct except:
    A. Orthodromic stimulation of the plantar sensory nerve produce a small amplitude response
    B. Bilateral absent plantar sensory responses in middle aged or older individuals have no significance
    C. In suspected tarsal tunnel syndrome, one side abnormal plantar response is diagnostic of tarsal tunnel syndrome

D. Medial and lateral plantar sensory potentials are unobtainable even in healthy subjects

12. **Small fiber peripheral neuropathy is seen in all of the following conditions except:**
    A. Diabetes
    B. Fabry disease
    C. Diphtheria
    D. Tangier disease

13. **The pattern of mononeuropathy multiplex is consistent with all of the following except:**
    A. Asymmetric involvement of the body
    B. Stepwise progression of individual nerves is commonly observed
    C. It affects larger nerves as opposed to small nerve twigs
    D. Cranial nerves are spared

14. **The electrodiagnostic findings in hereditary motor sensory neuropathy-type 1 include all of the following except:**
    A. Delayed latencies
    B. Slow conduction velocities with conduction block
    C. Slowing seen in children as young as 6 months of age
    D. Electromyographic (EMG) evidence of distal reinnervation with little spontaneous activity

15. **Hereditary motor sensory neuropathy-type 1 (HMSN-1), the most common hereditary neuropathy, is characterized by all of the following except:**
    A. Slowly progressive, distal, motor, and sensory demyelinating neuropathy
    B. Spares intrinsic foot and lower leg muscles
    C. Autosomal dominant
    D. *PMP22* gene defect

16. **In the electrodiagnostic evaluation of acute idiopathic demyelinating polyneuropathy (AIDP), at the end of 3 weeks after onset, a common constellation of nerve conduction findings includes the following except:**
    A. Median sensory response is abnormal
    B. Ulnar sensory response is abnormal
    C. Sural sensory response is abnormal
    D. Median motor response is abnormal

17. **In patients with radiculopathy, sensory nerve conduction studies typically show:**
    A. Absent sensory response
    B. Prolonged distal sensory latency
    C. Slow sensory conduction velocity
    D. Normal sensory conduction studies

18. An absent sural nerve sensory study with intact sensation over the lateral foot most likely represents:
    A. A chronic axonal lesion to the sural nerve
    B. A chronic demyelinating lesion to the sural nerve
    C. Collateral innervation of the lateral foot
    D. A technical problem

19. A patient presents with left hand tingling and numbness. The median motor conduction study is normal with normal left ulnar motor and sensory conduction study (see Fig. 13.2). The lesion is localized as:
    A. Left C6 radiculopathy
    B. Left carpal tunnel syndrome
    C. Left cubital tunnel syndrome
    D. Left ulnar sensory mononeuropathy

20. A patient presents with right hand numbness and tingling. The lesion is localized at:
    A. Right carpal tunnel syndrome
    B. Artifact
    C. Right Martin-Gruber anomaly
    D. Right median neuropathy above the elbow

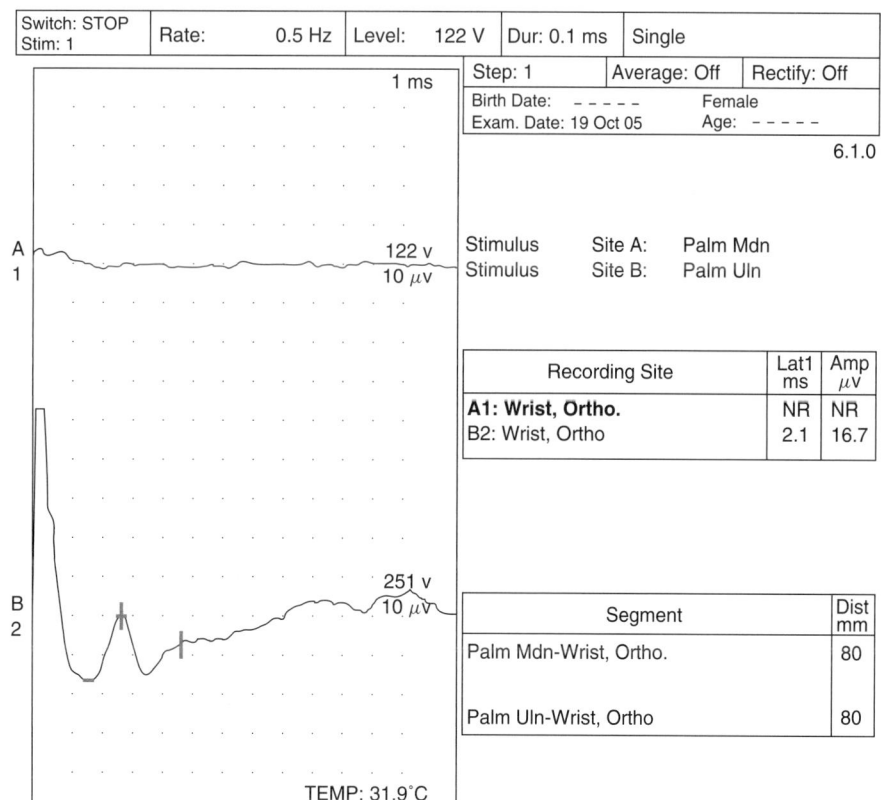

Figure 13.2. Median absent ulnar normal sensory

21. The sensory nerve action potential (SNAP) is normal, but the patient has clear sensory loss in the distribution of the sensory response being recorded. The reasons for these are all of the following except:
    A. Lesion proximal to the dorsal root ganglion
    B. Proximal demyelinating lesion
    C. Hyperacute axonal lesion
    D. Only large myelinated fibers are affected

ANSWERS

1. (B): The sensory responses are very small, and technical factors and electrical noise may distort the waveforms. For sensory conduction studies, the gain is usually set at 10 to 20 μV. The gain is set at microvolts; millivolts are used to measure motor conduction results. (**Preston and Shapiro 1998, p. 27**)

2. (C): The sensory nerves have a low threshold for stimulation compared to motor fibers and require a lower current (5 to 30 mA) to achieve supramaximal stimulation. The recording electrodes are placed in line over the nerve with recommended interelectrode distance of 3 to 4 cm. The active electrode is placed closest to the stimulator. Unlike motor studies, in sensory studies a conduction velocity can be calculated using one stimulation alone as the recording electrode is measuring a nerve action potential. (**Preston and Shapiro 1998, p. 29**)

3. (A): The antidromic technique results in stimulation of the entire mixed nerve including the motor fibers resulting in CMAP along with SNAP. This may become a problem when SNAP imbeds in CMAP. At times if SNAP is absent, one can misread CMAP as SNAP. The amplitude with antidromic technique is higher because the recording electrodes are closer to the nerve and is less subjected to noise or artifact. (**Preston and Shapiro 1998, p. 32**)

4. (B): The myelinated fibers conduct at a velocity of approximately 65 m/s, while the unmyelinated fibers conduct at a velocity of 1 to 2 m/s. Most fibers lie between these two extremes. (**Preston and Shapiro 1998, p. 37**)

5. (D): The drop in amplitude in proximal stimulation (conduction block) compared to the distal stimulation is seen with focal demyelinating neuropathies. Marked slowing of conduction velocity (slower than 75% of the lower limit of normal) is seen. Marked prolongation of distal latency (longer than 130% of upper limit of normal) is also a feature of demyelinating conditions. The CMAP amplitude can be normal or reduced in demyelinating neuropathies depending on the degree of distal demyelination or superimposed conduction block. (**Preston and Shapiro 1998, p. 39**)

6. (B): In patients with typical carpal tunnel syndrome (CTS), the median distal motor and sensory latencies, and minimum F-wave latencies are moderately to markedly prolonged. However, there is a group of patients with clinical symptoms and signs of carpal tunnel syndrome in whom these routine studies are normal (10% to 25%). In such patients, the electrodiagnosis of CTS may be missed. (**Preston and Shapiro 1998, p. 238**)

7. (**D**): A CTS study should show the median nerve orthodromic sensory latency delayed by >0.2 ms compared to the ulnar nerve. The median CMAP amplitude is usually much higher than ulnar. The picture depicts an orthodromic sensory response. The median nerve orthodromic sensory absolute latency >3 ms is usually abnormal, however, the absolute latency can be delayed because of multiple factors and therefore, absolute latency delay is usually not taken as a marker of CTS. (**Preston and Shapiro 1998, p. 238**)

8. (**D**): In comparison studies, identical distances are used between the stimulator and recording electrodes for median and ulnar nerves. These techniques create an ideal internal control in which several variables that are known to affect conduction time are held constant, for example, distance, temperature, age, and nerve size. If the distal motor and sensory latencies are normal then comparison studies often are very helpful because they rely on patient's own nerve rather than on population normal values. (**Preston and Shapiro 1998, p. 238**)

9. (**D**): In comparison studies, very small difference (0.4 ms) between median and ulnar nerve is considered abnormal. Attention must be paid to all technical factors to avoid false-positive results. Technical factors such as distance measurement, stimulus artifact, supramaximal stimulation, and electrode placement can all affect the results. It is essential to avoid over stimulation because it can cause unintentional stimulus spread to an adjacent nerve. (**Preston and Shapiro 1998, p. 238**)

10. (**A**): The absence of a sural sensory response is not seen in tarsal tunnel syndrome. Tarsal tunnel syndrome is the compression of distal tibial nerve under the flexor retinaculum on the medial side of ankle. This results in a delayed or absent response in the medial and lateral plantar nerve responses. The absence of sural sensory response indicates a lesion above the tarsal tunnel syndrome and can be affected by a sciatic nerve lesion, lumbosacral plexopathy, and peripheral neuropathy. (**Preston and Shapiro 1998, p. 331**)

11. (**A**): Evaluation of tarsal tunnel syndrome is greatly simplified if one side is abnormal as the other side acts as a control (Fig. 13.3). Surface plantar sensory studies are difficult to perform in healthy subjects. Orthodromic stimulation of the plantar sensory nerve produces very small amplitude responses, making it necessary to average many potentials. Medial and lateral plantar sensory potentials are unobtainable even in healthy subjects and similarly bilateral absent plantar sensory responses in middle-aged or older individuals have no significance. (**Preston and Shapiro 1998, p. 331**)

12. (**C**): Only a few peripheral neuropathies preferentially affect small fibers. Diabetes, Fabry disease, and Tangier disease all cause small fiber peripheral neuropathy while diphtheria does not typically affect the small fibers. Small fiber neuropathy manifests as an autonomic dysfunction and a distal peripheral sensory deficit for pinprick and cold, often associated with painful paresthesia. Routine

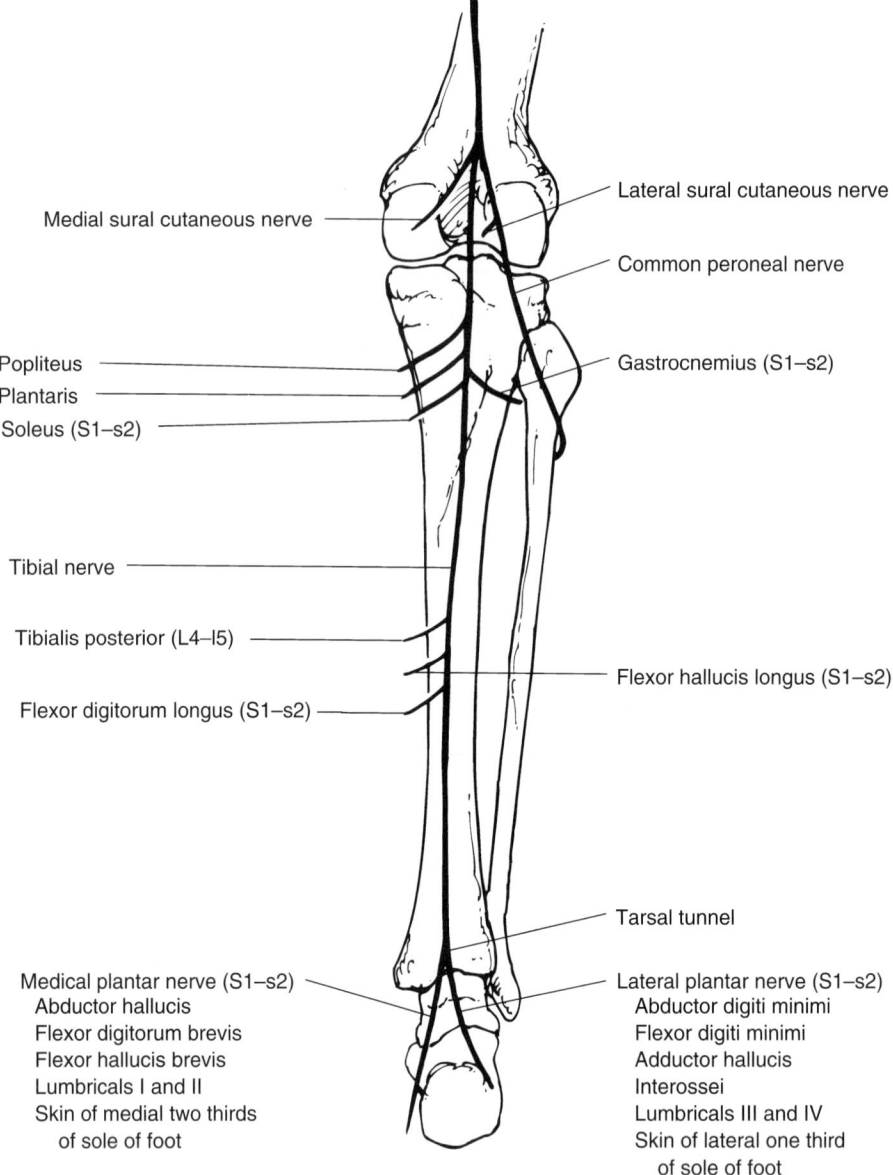

Figure 13.3. The tibial nerve (From Brazis PW, Masdeu JC, Biller J. *Localization in clinical neurology, 5th ed.* Philadelphia: Lippincott Williams & Wilkins; 2007)

nerve conduction studies assess only large myelinated fibers and can be completely normal in small fiber neuropathies. (**Preston and Shapiro 1998, p. 356**)

13. (D): Mononeuropathy multiplex is a distinctive pattern presenting as asymmetric, step-wise progression of individual peripheral and cranial nerves usually of the large nerves and not of small nerve twigs. As time passes, a confluent pattern may develop that is often difficult to distinguish from a diffuse polyneuropathy. Mononeuropathy multiplex is most commonly seen in the setting of vasculitic neuropathy. (**Preston and Shapiro 1998, p. 358**)

14. (B): In a patient with inherited demyelinating polyneuropathies, all myelin tends to be affected equally, resulting in uniform slowing of conduction velocity.

Because of this pathophysiology, conduction block is not seen in inherited polyneuropathies such as HMSN-1. Nerve conduction studies show marked slowing of conduction velocity (below 75% of the lower limit) and markedly prolonged distal latencies. Prominent slowing is often recorded during the first 3 to 5 years of life. Slowing has been documented in patients as young as 6 months of age. The EMG typically shows evidence of distal reinnervation with little spontaneous activity. (**Preston and Shapiro 1998, p. 365**)

15. (**B**): HMSN-1 is the most common hereditary neuropathy in clinical practice. It is a slowly progressive, distal, motor and sensory demyelinating neuropathy. It affects the intrinsic foot and lower leg muscles and is associated with pes cavus, hammertoes, and hypertrophic nerves. Sensory symptoms are less common than motor complaints. There are no cranial nerve signs. The inheritance is most commonly autosomal dominant and the gene association is *PMP22*. There are also many families in whom the inheritance pattern is autosomal recessive or X-linked recessive. It is the same gene that causes pressure palsy. DNA testing for *PMP22* is now commercially available. (**Preston and Shapiro 1998, p. 366**)

16. (**C**): Like the results of the motor studies, all sensory studies are normal in the beginning of AIDP. However, at the end of the first week sural sparing may be seen while the ulnar, superficial peroneal sensory and median sensory responses are abnormal. This happens because of preferential involvement of smaller myelinated fibers early in the AIDP. Some people believe that sural sparing in the presence of a typical clinical picture is virtually diagnostic of AIDP. The sural fibers are larger and more myelinated than the median and ulnar nerves. Median and ulnar sensory responses are diminished in the digits where they have tapered and have a less myelin. Median motor response, especially F-wave response is usually abnormal by this time. (**Preston and Shapiro 1998, p. 383**)

17. (**D**): Sensory studies are the most important part of the nerve conduction studies in the evaluation of radiculopathy. The sensory nerve action potential remains normal in lesions proximal to the dorsal root ganglion. Nearly all radiculopathies damage the root proximal to the dorsal root ganglion. Conversely, lesions distal to the dorsal root ganglion (plexopathy and peripheral nerve) result in decreased SNAP amplitude. The presence of a normal SNAP in the same distribution as sensory symptoms should always suggest a lesion proximal to the dorsal root ganglion (radiculopathy). (**Preston and Shapiro 1998, p. 418**)

18. (**D**): Technical problems can easily lead to abnormal or absent responses. The key in determining the difference between a technical abnormality and a true lesion is the lack of clinical electrophysiologic correlation. The study is best done with patient lying on his side. Side-to-side comparison of amplitude and latency are often helpful. The examiner should investigate possible technical causes and should retest the nerve several times before declaring the result abnormal. (**Preston and Shapiro 1998, p. 134**)

19. (**B**): Figure 13.1 shows an absent left median sensory response while the ulnar sensory response is normal. This effectively rules out radiculopathy as with root lesions, sensory conduction studies are normal. Left ulnar sensory response rules out left cubital tunnel syndrome or ulnar sensory mononeuropathy. The only

other reasonable option is left carpal tunnel syndrome. However, orthodromic sensory studies are susceptible to technical problems. It is imperative to do median antidromic sensory conduction study to confirm the sensory mononeuropathy. Antidromic sensory studies are more sensitive and are relatively resistant to technical problems. **(Preston and Shapiro 1998, p. 419)**

20. (A): Figure 13.2A shows a delayed median antidromic sensory latency compared to radial sensory latency. Figure 13.2B shows a prolonged median distal motor latency and absence of conduction block. These findings are consistent with a right carpal tunnel syndrome. Martin-Gruber anomaly is the most commonly encountered anomaly in the upper extremity with a cross over of median-to-ulnar fibers. The Martin-Gruber anastomosis (MGA) is seen in approximately 15% to 30% of patients. The most common MGA anomaly shows decreased ulnar CMAP amplitude (recording at first dorsal interosseous) with below elbow stimulation compared to wrist stimulation. This is not seen in Figure 13.2A, hence it is not a correct option. **(Preston and Shapiro 1998, p. 78)**

21. (D): Sensory responses may remain normal on conduction studies in three scenarios. (a) If the dorsal root ganglion and peripheral nerve are intact and the lesion is proximal to the dorsal root ganglion, for example, myelopathy or radiculopathy, (b) similarly with pure demyelination, underlying axon remain intact and wallerian degeneration never occurs. Therefore, the distal nerve continues to conduct normally, and (c) similarly, after an axonal lesion the distal nerve continues to work normally for first several days before Wallerian degeneration occurs. During this time, the distal nerve conduction studies are normal. The routine nerve conduction studies test the large myelinated fibers, but not the small fibers. Hence in small fiber, neuropathy nerve conduction study can be completely normal. **(Preston and Shapiro 1998, p. 226)**

References

1. Preston DC, Shapiro BE. *Electromyography and neuromuscular disorders*, 1st ed. Butterworth-Heinemann; 1998.

CHAPTER 14

Motor Nerve Conduction Studies

QUESTIONS

1. **All the statements regarding F-wave are correct except:**
   A. F-wave response is actually a small compound muscle action potential (CMAP) representing 1% to 5% of the muscle fibers
   B. F-wave is a motor response
   C. F-wave represents a reflex arc
   D. F-wave may be absent in sleeping or sedated patients

2. **Normal F-wave latency in the lower extremity is:**
   A. 25 to 32 ms
   B. 35 to 44 ms
   C. 45 to 56 ms
   D. 60 to 70 ms

3. **F-wave may have its greatest usefulness in identifying:**
   A. Myasthenia gravis
   B. Amyotrophic lateral sclerosis (ALS)
   C. Diabetic sensory neuropathy
   D. Guillain-Barré syndrome (GBS)

4. **All statements regarding H reflex are correct except:**
   A. H reflex is a true reflex with a sensory afferent segment, a synapse, and a motor efferent segment
   B. The typical H reflex latency is approximately 30 ms
   C. Supramaximal stimulation is used to elicit an H reflex response
   D. H reflex interside difference of >1.5 ms is considered significant

143

5. All statements regarding repetitive nerve stimulation (RNS) are correct except:
   A. The recording electrode's movement in relationship to the muscle movement may change the CMAP
   B. Submaximal stimulation can create a CMAP decrement
   C. CMAP decrement is accentuated if the limb is cold
   D. Pyridostigmine needs to be stopped 3 to 4 hours before the RNS

6. Optimal frequency for RNS to diagnosis of myasthenia gravis is:
   A. 1 Hz
   B. 3 Hz
   C. 10 Hz
   D. 50 Hz

7. With 3 Hz RNS, the lowest CMAP (maximal decrement) is usually seen with:
   A. First stimulation
   B. Third stimulation
   C. Fifth stimulation
   D. Seventh stimulation

8. Routine nerve conduction studies (NCSs) are normal in patients with clinical symptoms and signs of carpal tunnel syndrome (CTS) in ___% of patients.
   A. 5%
   B. 20%
   C. 30%
   D. 40%

9. Median versus ulnar comparison tests are good tests to confirm CTS for following reasons except:
   A. It creates an ideal internal control
   B. Distance is different
   C. Temperature is constant
   D. Nerve fibers size is constant

10. All statements regarding median versus ulnar F-wave comparison is correct except:
    A. Median F-wave latency is normally 1 to 2 ms delayed than the ulnar F-wave latency
    B. Median F-wave latency is normally 1 to 2 ms shorter than the ulnar F-wave latency
    C. Median F-wave latency is usually the same as ulnar F-wave latency
    D. Abnormal median F-wave latency alone is diagnostic of CTS

11. A proximal median neuropathy is differentiated from CTS by which of the following tests?
    A. Median F-wave latency
    B. Acute denervation pattern in abductor policis brevis
    C. Acute denervation pattern in pronator teres
    D. Acute denervation pattern in the flexor carpi ulnaris

12. **The electrophysiologic marker of a conduction block in the ulnar nerve at the elbow is:**
    A. A drop in CMAP amplitude of 5%
    B. A drop in CMAP latency of 20%
    C. Dispersion of 10% in the CMAP waveform on stimulation above the elbow
    D. An abrupt drop in conduction velocity at the elbow

13. **All statements regarding the routine ulnar NCS are correct except:**
    A. When stimulating at below-elbow site, the stimulator should be at least 3 to 4 cm distal to the groove to ensure a stimulation point distal to the cubital tunnel
    B. Distance between the below-elbow site to above-elbow site should be >15 cm
    C. A higher current is often required to ensure supramaximal stimulation of the ulnar nerve at the below-elbow site
    D. Distance between below-elbow site to above-elbow site should not be <10 cm

14. **All statements about an ulnar neuropathy are correct except:**
    A. There are no ulnar innervated muscles above the elbow
    B. If all ulnar innervated muscles are abnormal and there is no conduction block then it cannot localize the lesion
    C. If all ulnar innervated muscles are abnormal and there is a conduction block at the elbow then it localizes the lesion to the elbow
    D. If both ulnar motor and sensory nerve conduction (SNC) studies are abnormal then it localizes the lesion at the wrist

15. **Electromyographic (EMG) and NCS findings that help to differentiate a C8-T1 radiculopathy from an ulnar neuropathy at the elbow are:**
    A. Low median CMAP amplitude
    B. Low ulnar CMAP amplitude
    C. Acute denervation in flexor carpi ulnaris
    D. Acute denervation in flexor digitorum profundus

16. **All statements regarding radial neuropathy are correct except:**
    A. The superficial radial sensory nerve is easy to stimulate and record
    B. Most cases of posterior interosseous neuropathy are pure demyelinating in nature and a conduction block is demonstrated
    C. Radial neuropathy at the spiral groove shows a conduction block with stimulation of the radial nerve proximal to the spiral groove
    D. Posterior interosseous neuropathy results in a normal superficial radial sensory nerve action potential (SNAP)

17. **The pattern of mononeuropathy multiplex is consistent with all of the following except:**
    A. Asymmetric involvement of the body
    B. Stepwise progression of individual nerve is commonly observed
    C. It affects larger nerves as opposed to small nerve twigs
    D. Cranial nerves are spared

18. **All statements regarding acute inflammatory demyelinating polyradiculoneuropathy (AIDP) are correct except:**
    A. Most patients will have motor abnormalities
    B. Early in the illness, the EMG shows normal motor unit action potentials (MUAPs) with decreased recruitment in weak muscles
    C. Fibrillations develop 5 to 6 weeks
    D. Prognosis: low distal amplitude is a prognostic measure (0% to 20% lower limit at 3 to 5 weeks) for delayed and incomplete recovery

19. **Chronic inflammatory demyelinating polyneuropathy EMG/NCS findings are all of the following except:**
    A. Secondary axonal involvement seen
    B. Is a multifocal process
    C. Conduction block is seen
    D. Paraspinal muscles are spared

20. **The following statements regarding multifocal motor neuropathy with conduction block are correct except:**
    A. Is a pure motor neuropathy
    B. Progressive, asymmetric weakness and wasting more distally than proximally
    C. May mimic ALS
    D. Patients are older than 50 years

21. **In the electrodiagnostic evaluation of AIDP, early in the disease, all the following findings are seen except:**
    A. Median distal motor latency is normal
    B. Median conduction velocity is normal
    C. Median sensory response is normal
    D. Median F-wave latency is normal

22. **Electrophysiologic criteria for AIDP is consistent with all of the following choices except:**
    A. Prolonged distal latencies in two or more nerves that exceed 115% of the upper limit normal
    B. Conduction velocity slowing in two or more nerves <90% of the lower limit normal
    C. Prolonged F-wave latency in one or more nerves >110% of the upper limit normal
    D. Conduction block or temporal dispersion in one or more nerves

23. **Electrophysiologic criteria for chronic inflammatory demyelinating polyneuropathy (CIDP) is consistent with all except:**
    A. Prolonged distal latencies in two or more nerves >130% of upper limit normal
    B. Conduction velocity slowing in two or more nerves <75% of the lower limit normal

C. Prolonged F-wave latency in one or more nerves >110% of the upper limit normal
D. Conduction block or temporal dispersion in one or more nerves

24. **The tracings in Figure 14.1 of a patient with right hand pain suggest the lesion to be at:**
    A. Right lower trunk of the brachial plexus
    B. Right median mononeuropathy distal to carpal tunnel
    C. Right CTS
    D. Right C8 radiculopathy

25. **Patient presents with diplopia and difficulty swallowing. Routine NCSs are normal. Right ulnar nerve repetitive stimulation test and single fiber electromyographic (SFEMG) tracings are shown in Figure 14.2. Final diagnosis is:**
    A. Artifact
    B. Abnormal study consistent with myasthenia gravis
    C. Normal study
    D. No sufficient data

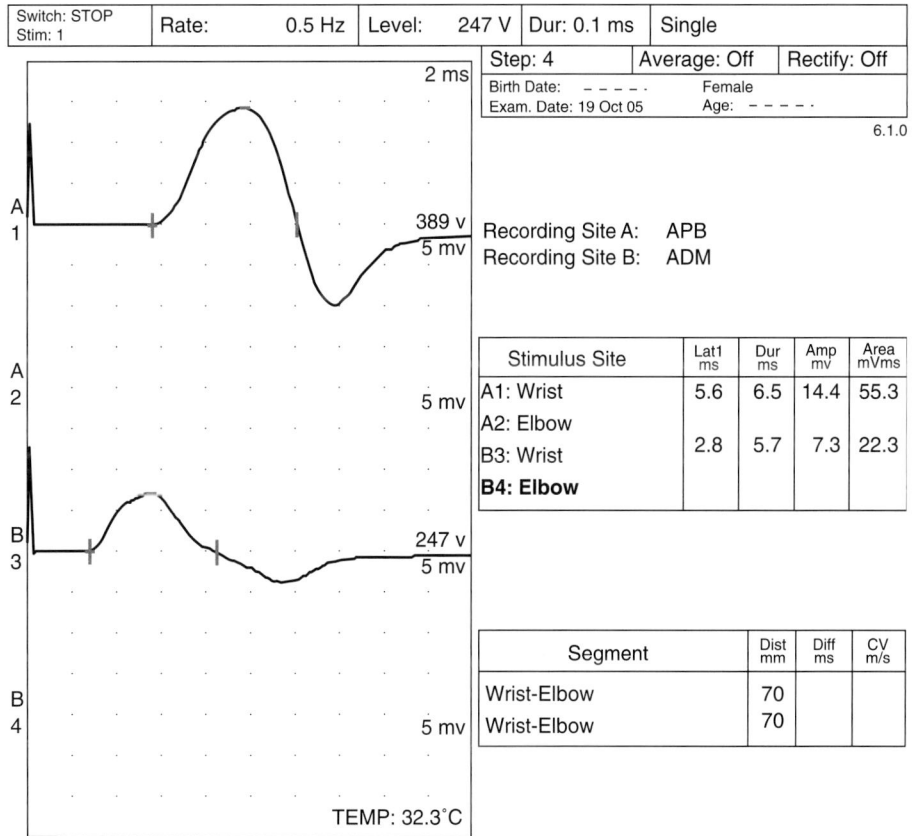

Figure 14.1. Median versus ulnar motor nerve conduction (MNC) and median versus ulnar sensory nerve conduction (SNC)

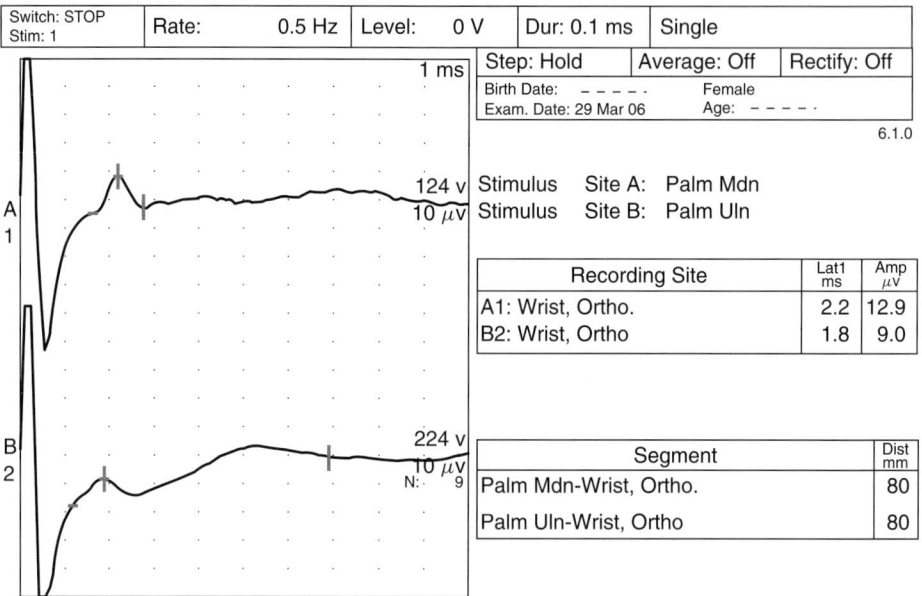

Figure 14.1. (Continued)

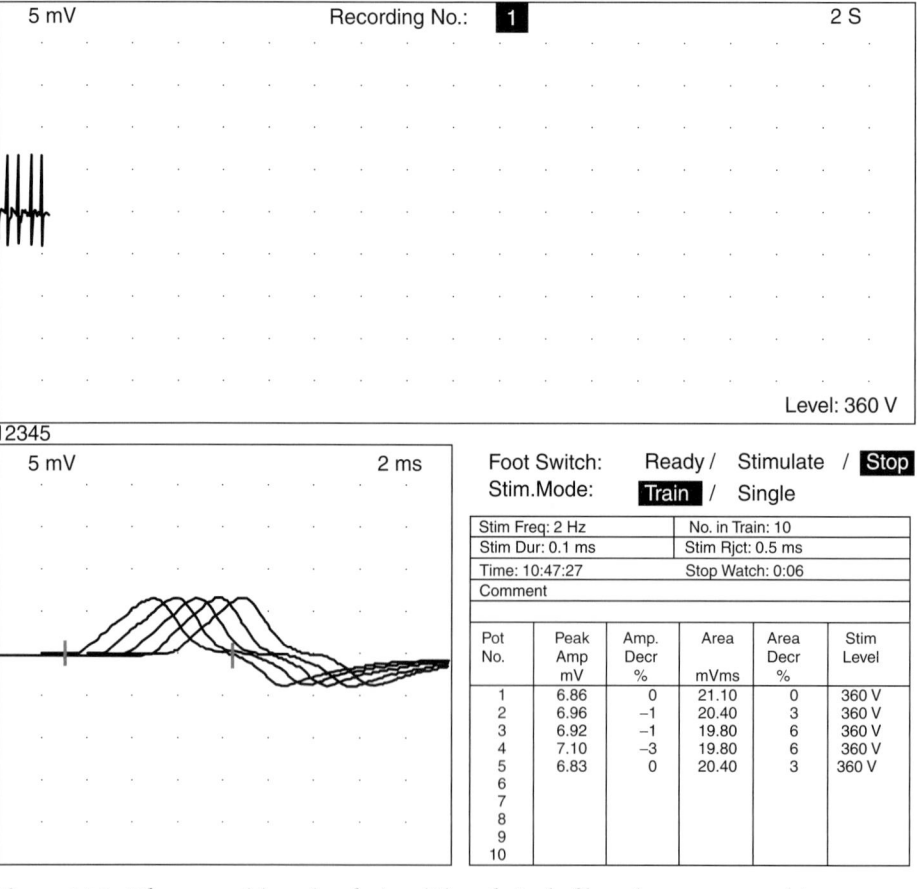

Figure 14.2. Ulnar repetitive stimulation (**A**) and single fiber electromyographic (SFEMG) tracing (**B**)

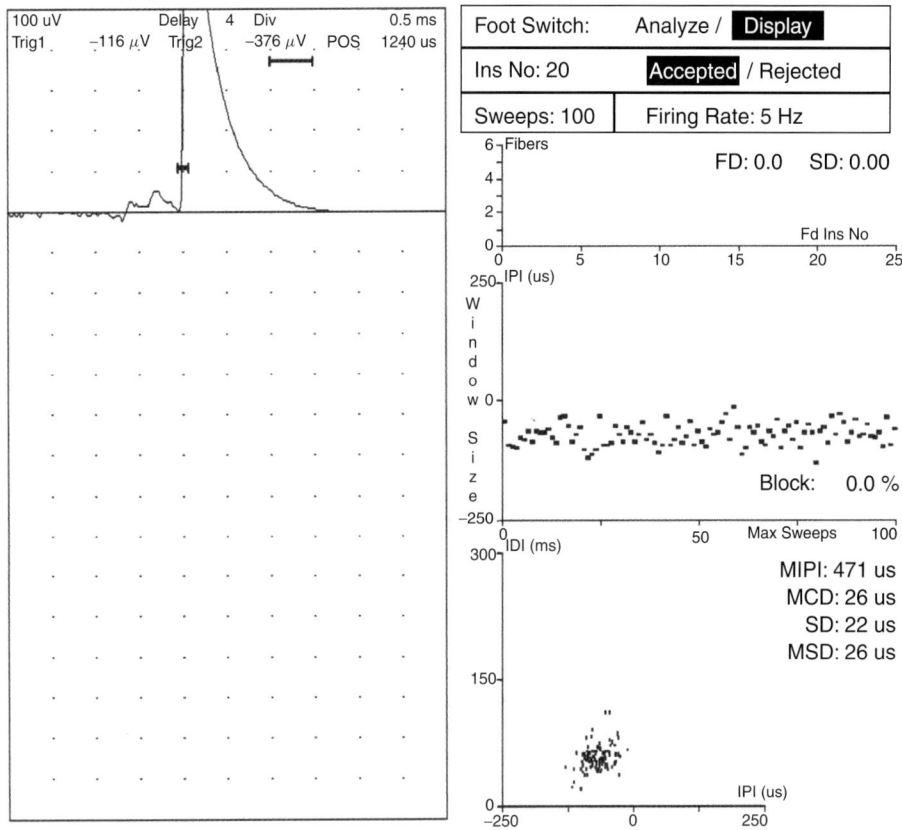

Figure 14.2. (Continued)

26. A patient presents with left hand pain and tingling. The median motor nerve conduction (MNC) and SNC studies are normal. The ulnar NCS tracing shown in Figure 14.3 indicates:
    A. Left cubital tunnel syndrome
    B. Ulnar mononeuropathy, nonlocalizable
    C. Left ulnar mononeuropathy in the forearm
    D. Left brachial plexopathy

27. The tracing in Figure 14.4 is of a patient with pain and numbness in the ulnar aspect of the hand. The inching techniques in ulnar nerve evaluation is preferred because:
    A. Any abrupt increase in latency or drop in amplitude between successive stimulation sites implies focal demyelination
    B. The latency difference of >0.2 ms is abnormal
    C. This is a very easy technique
    D. It does not add much except academic satisfaction

150 *Section 4:* Nerve Conduction Studies

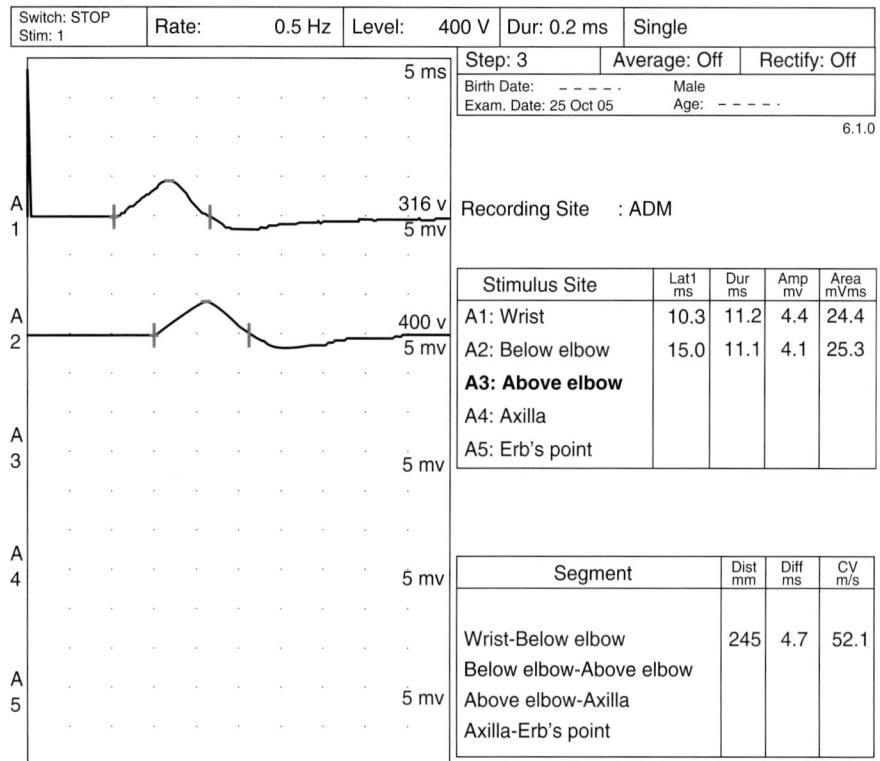

Figure 14.3. Ulnar motor nerve conduction (MNC) demyelinating neuropathy

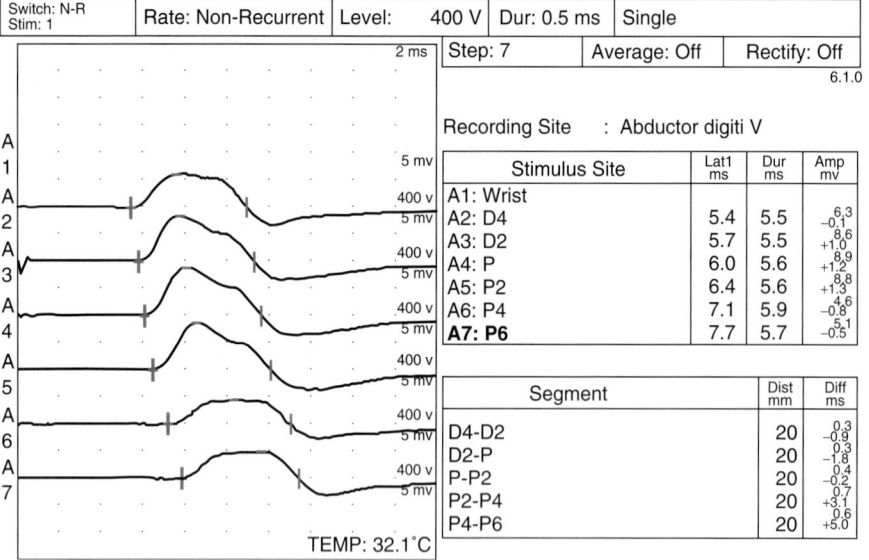

Figure 14.4. Ulnar inching

28. **The F-wave response as seen in Figure 14.5 is normal. Choose the correct response related to F-waves:**
   A. If the stimulator is moved proximally the latency of F-wave response increases
   B. The F-wave represents the reflex arc involving the sensory nerve dorsal horn, interneurons, anterior horn cell, and the motor nerve

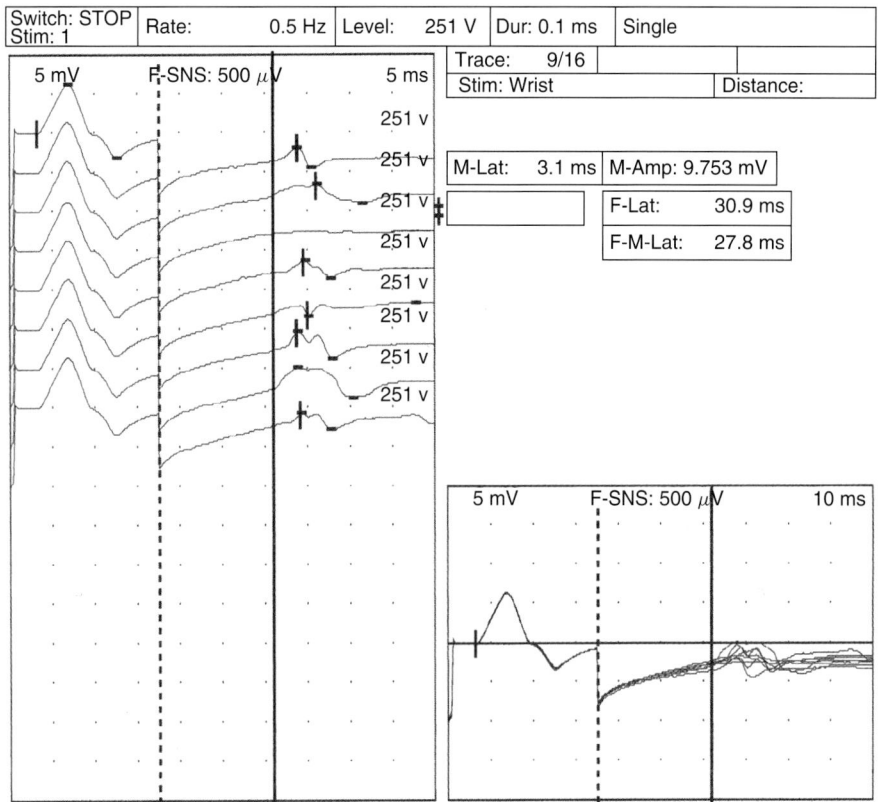

Figure 14.5. Ulnar F-wave

C. F-wave response varies slightly in latency, configuration, and amplitude because a different population of anterior horn cells is activated with each stimulation
D. Normal F-wave latency for ulnar nerve is 40 ms

29. Patient presents with right foot drop after having mitral valve replacement surgery. The right peroneal motor response is shown in Figure 14.6. All statements below are correct except:
    A. Peroneal distal motor latency is normal
    B. There is significant conduction block seen across the fibular head
    C. The response at the fibular head shows temporal dispersion
    D. The CMAP amplitude is abnormal and low

30. Blink reflex abnormalities can be seen with all except:
    A. GBS
    B. Medullary lesion
    C. Midbrain lesion
    D. Axonal neuropathy

31. Abnormal temporal dispersion with MNC studies is consistent with:
    A. Neuromuscular junction disorder
    B. Demyelinating disorder
    C. Temperature effect
    D. Amplifier turned off

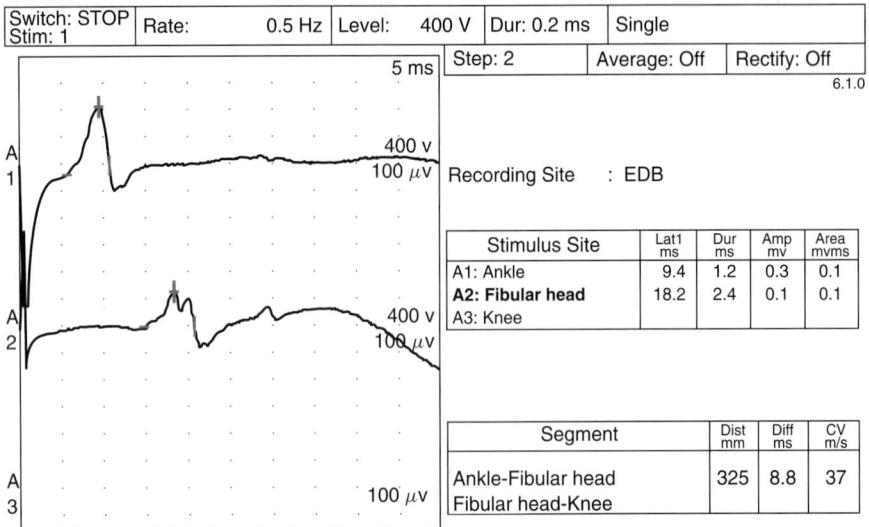

Figure 14.6. Peroneal block at fibular head

32. A 45-year-old lady presents with left hand weakness. You perform the median and ulnar distal motor latency testing. The response is shown in Figure 14.7. The diagnosis is consistent with:
    A. Left cubital tunnel syndrome
    B. Left CTS
    C. Left myasthenia gravis
    D. Left cervical radiculopathy

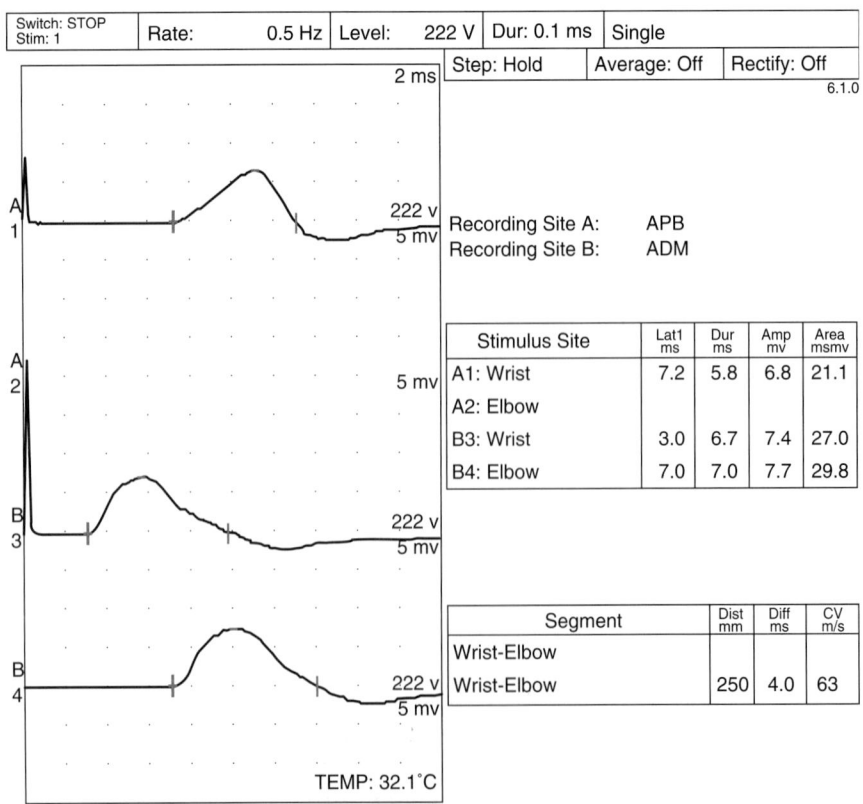

Figure 14.7. Median versus ulnar motor nerve conduction (MNC)

ANSWERS

1. (C): F-wave is a pure motor response and represents a small CMAP representing 1% to 5% of the muscle fibers. Normal F-wave persistence is usually 80% to 100% but always >50%. F-wave can be obtained from any motor nerve. It is best obtained with distal stimulation. F-wave may be absent in sleeping or sedated patients and in some healthy patients particularly in the peroneal nerve. Supramaximal stimulation should be used with stimulator turned around so that the cathode is more proximal. One can record equally good F-waves without reversing the cathode and anode but the possibility of anodal block increases. (**Preston and Shapiro 1998, p. 45**)

2. (C): F-wave can be obtained from any motor nerve in lower extremity. It is best obtained with distal stimulation as proximal stimulation can result in superimposition of the F-waves on the terminal CMAP. Supramaximal stimulation must always be used and it is best to reverse the cathode and anode to eliminate the possibility of anodal block. F-wave latency is longer in tall patients. Mildly prolonged F-waves in tall patients should not interpreted as abnormal. (**Preston and Shapiro 1998, p. 45**)

3. (D): F-wave responses may have their greatest usefulness in identifying early polyradiculopathy such as occurs in GBS. GBS commonly begins with demyelination of the nerve roots before the distal nerve segments. Routine motor nerve studies may be normal in early GBS while F-wave prolongation may be the only sign of abnormality. Myasthenia gravis is a neuromuscular disease and F-wave response will not help to diagnose it. Diabetic sensory neuropathy will not affect the F-wave response because the F-wave is recorded solely from motor fibers. ALS may have prolonged or absent F-waves in later stage of the disease. However F-waves are rarely useful in establishing the diagnosis of ALS as the needle examination plays a much greater role. (**Preston and Shapiro 1998, p. 50**)

4. (C): H reflex is a true reflex with a sensory afferent (Ia sensory fibers) segment, a synapse, and a motor efferent segment. The typical H reflex latency is approximately 30 ms. Comparison with the contralateral side is more useful in assessing a unilateral lesion; any difference of >1.5 ms is considered significant. A supramaximal response will abolish the H reflex and commonly lead to the elicitation of an F-wave. Therefore, a slowly increasing submaximal stimulation is used to achieve H reflex. (**Preston and Shapiro 1998, p. 54**)

5. (C): Dislodging or movement of the recording electrode in relationship to the muscle movement may change the CMAP. RNS is obtained by supramaximal stimulation as a submaximal stimulation can create a CMAP decrement which can simulate a decrement and make the test false positive. Limb temperature can affect the test resulting in a false-negative study. CMAP decrement is diminished if the limb is cold. Pyridostigmine needs to be stopped 3 to 4 hours before the RNS as it makes more acetylcholine available to bind at the acetylcholine receptors and may potentially diminish CMAP decrements, resulting in a normal study or a decrement insufficient to establish a neuromuscular junction defect. (**Preston and Shapiro 1998, p. 69**)

6. (B): The optimal frequency for RNS is 2 or 3 Hz. This frequency is low enough to prevent calcium accumulation, but high enough to deplete the quanta in the immediately available store before the mobilization store starts to replenish it. For rapid RNS the optimal frequency is 30 to 50 Hz. **(Preston and Shapiro 1998, p. 70)**

7. (C): The lowest CMAP is usually the third or fourth stimulation. By the fifth or sixth stimulation the decrement begins to improve because the mobilization store of acetylcholine begins to resupply the immediately available store. Any decrement of >10% is defined as abnormal. Healthy subjects should have no decrement. **(Preston and Shapiro 1998, p. 71)**

8. (B): In patients with typical CTS, the median distal motor and sensory latencies and minimum F-wave latencies are moderately to markedly prolonged. However there is a group of patients with clinical symptoms and signs of CTS in whom these routine studies are normal (10% to 25%) In such patients the electrodiagnosis of CTS may be missed. **(Preston and Shapiro 1998, p. 238)**

9. (D): In comparison studies identical distances are used between the stimulator and recording electrodes for median and ulnar nerves. These techniques create an ideal internal control in which several variables that are known to affect conduction time are held constant, for example, distance, temperature, age, and nerve size. If the distal motor and sensory latencies are normal then comparison studies often are very helpful as they rely on patient's own nerve rather than population normal values. **(Preston and Shapiro 1998, p. 238)**

10. (B): In normal individuals, the minimum F-wave latency from the median nerve is approximately 1 to 2 ms shorter than the minimum latency from the ulnar nerve. A reversal of that pattern is considered abnormal. The F-wave latency comparison is a nonspecific test because F-wave measures conduction along the entire length of nerve and cannot localize the lesion. Therefore it is used in conjunction with other sensitive tests. **(Preston and Shapiro 1998, p. 245)**

11. (C): Median F-wave latency assesses the whole length of the median nerve. However F-wave latency will be abnormal in both CTS and proximal median neuropathy. Similarly abductor policis will be affected in both conditions. Flexor carpi ulnaris is innervated by the ulnar nerve and will not be affected by any median neuropathy. Proximal median neuropathy at or proximal to the branch to pronator teres will result in acute denervation in the pronator teres muscle. **(Preston and Shapiro 1998, p. 262)**

12. (D): Ulnar motor studies in healthy subjects have shown a maximum drop in CMAP amplitude of 10% comparing below and above groove stimulation. Accordingly any drop in amplitude of >10% from below to above the groove is consistent with true conduction block. Minimal dispersion can be seen in healthy subjects and will not qualify for a conduction block. A drop in CMAP latency of 20% is significant but does not qualify as conduction block. **(Preston and Shapiro 1998, p. 273)**

13. (B): When stimulating at below-elbow site, the stimulator should be at least 3 to 4 cm distal to the groove (because the cubital tunnel length can be variable) to ensure a stimulation point distal to the cubital tunnel. The distance between below-elbow site to above-elbow site should not be <10 cm. If a short distance is used, slight errors in measurement may create large differences in calculated conduction velocities. Because the ulnar nerve is deep to the flexor carpi ulnaris at the below-elbow site, higher current is often required to ensure supramaximal stimulation of the ulnar nerve at this site. (**Preston and Shapiro 1998, p. 276**)

14. (D): If both ulnar motor and SNC studies are abnormal then it localizes the lesion at the elbow because the dorsal ulnar cutaneous sensory branch exits 5 to 8 cm proximal to the wrist to supply the dorsal medial hand. A lesion at the wrist will not affect the sensation in the medial hand. If all ulnar innervated muscles are abnormal and there is a conduction block at the elbow then it localizes the lesion at the elbow. If there is no conduction block then the ulnar neuropathy can be localized roughly, if the needle examination is abnormal to above the abnormal muscles. (**Preston and Shapiro 1998, p. 277**)

15. (A): C8-T1 radiculopathy will affect flexor carpi ulnaris and flexor digitorum profundus muscles which are ulnar innervated muscles and will not help in differentiating the two conditions. C8 fibers are also in the median nerve and will result in a low median CMAP amplitude. However, median CMAP amplitude will not be affected by ulnar neuropathy at the elbow. (**Preston and Shapiro 1998, p. 271**)

16. (B): In cases of radial neuropathy at the spiral groove, conduction block is seen with stimulation of the radial nerve proximal to the spiral groove. The relative drop in proximal to distal CMAP amplitude gives some indication of the proportion of fibers blocked. Most cases of posterior interosseous neuropathy are purely axonal in nature and a conduction block is usually not seen. A normal superficial radial sensory response is seen in posterior interosseous neuropathy because the superficial radial sensory nerve comes off before the posterior interosseous nerve. (**Preston and Shapiro 1998, p. 291**)

17. (D): Mononeuropathy multiplex is a distinctive pattern presenting as asymmetric, step-wise progression of individual peripheral and cranial nerves usually of the large nerves and not of small nerve twigs. As time passes, a confluent pattern may develop that is often difficult to distinguish from a diffuse polyneuropathy. Mononeuropathy multiplex is most commonly seen in the setting of vasculitic neuropathy. (**Preston and Shapiro 1998, p. 273**)

18. (C): Most patients with AIDP will have weakness as the predominant feature. As in most neuropathies affecting the axon, fibrillation potentials do not develop until 3 to 4 weeks after the onset of the illness. Early in the illness, the only EMG abnormality will be reduced recruitment of MUAPs that are normal in morphology. The prognosis of AIDP is worse in patients with low CMAPS, in those whose therapy with intravenous immunoglobulin (IVIG) or plasma exchange has been delayed >7 days, in patients with respiratory involvement, and in elderly patients. (**Preston and Shapiro 1998, p. 362**)

19. (D): CIDP on NCS typically shows markedly prolonged distal latencies (>130% of the upper limit), slow conduction velocity (<75% of the lower limit) and prolonged F-wave (>130% of the upper limit). It is a multifocal process and conduction block is seen. Secondary axonal involvement is the rule. Needle EMG shows evidence of chronic and ongoing axonal loss. Because CIDP is actually a polyradiculopathy rather than a polyneuropathy alone, changes are often also seen in the paraspinal muscles. (**Preston and Shapiro 1998, p. 367**)

20. (**D**): Multifocal motor neuropathy with conduction block presents with progressive, asymmetric weakness, and wasting more distally than proximally. Most patients are younger than 50 years. There is a male predominance. Occasional patients have prominent weakness without wasting (consistent with pure demyelination). It may mimic ALS but bulbar function and sensation are spared. This condition is associated with anti-GM1 antibodies in 50% to 80% of patients. Asymmetry in presentation, upper extremity predominance, and lack of response to steroids help differentiate this condition from CIDP. (**Preston and Shapiro 1998, p. 367**)

21. (**D**): In AIDP the earliest finding on electrodiagnostic testing is the absent or delayed F-wave latency. The distal motor latencies, conduction velocity, and sensory responses are normal in the beginning of the disease. Later routine motor studies show prolonged distal latencies and conduction block. These changes are present in 50% of patients by 2 weeks and in 85% by 3 weeks. Occasionally some axonal changes are seen on the NCSs by this time. Few patients with AIDP have SNC abnormalities early in the illness. (**Preston and Shapiro 1998, p. 383**)

22. (**C**): Electrophysiologic criteria for AIDP is helpful in confirming the diagnosis of AIDP. This criteria indicates that prolonged distal latencies in two or more nerves >115% of upper limit normal for normal CMAP amplitude and >125% upper limit normal if amplitudes are low. The other criteria is conduction velocity slowing in two or more nerves <90% of the lower limit normal or <80% lower limit if amplitude is low. Prolonged F-wave latency in one or more nerves is seen >125% of the upper limit normal. Conduction block or temporal dispersion is seen in one or more nerves. (**Preston and Shapiro 1998, p. 365**)

23. (**C**): Electrophysiologic criteria for CIDP is helpful in confirming the diagnosis of CIDP. This criteria indicates that prolonged distal latencies in two or more nerves >130% of upper limit normal for normal CMAP amplitude. The other criterion is conduction velocity slowing in two or more nerves <75% of the lower limit. Prolonged F-wave latency in one or more nerves is seen >130% of the upper limit normal. Conduction block or temporal dispersion is seen in one or more nerves as unequivocal conduction block or temporal dispersion. (**Preston and Shapiro 1998, p. 365**)

24. (**B**): The tracing in Figure 14.1 shows median distal motor latency was significantly prolonged compared to the ulnar nerve and median amplitude was low indicating secondary axonal involvement of the median motor fibers. The normal median sensory response indicates that the lesion is distal to the carpal tunnel. Distal to the carpal tunnel, the median nerve divides into sensory and motor branches. Digital sensory branches supply the fingers while the motor branches supply the first and second lumbrical and the recurrent thenar motor branch supply the thenar muscles. (**Preston and Shapiro 1998, p. 233**)

25. (C): The tracing in Figure 14.2 shows ulnar RNS without a significant decrement in the CMAP. The optimal frequency for RNS is 2 or 3 Hz. This frequency is low enough to prevent calcium accumulation, but high enough to deplete the quanta in the immediately available store before the mobilization store starts to replenish it. The lowest CMAP is usually the third or fourth stimulation. By the fifth or sixth stimulation the decrement begins to improve because the mobilization store of acetylcholine begins to resupply the immediately available store. Any decrement of >10% is defined as abnormal. Healthy subjects should have no decrement as was in this case. The SFEMG tracing is normal without significant jitter or blocking. (**Preston and Shapiro 1998, p. 70**)

26. (B): The tracing in Figure 14.3 shows ulnar motor response at wrist and at elbow. The ulnar distal motor latency is severely delayed with decreased amplitude. This is consistent with ulnar demyelinating ulnar motor neuropathy with secondary axonal involvement (indicated by low amplitude). There is no conduction block seen across the elbow. This effectively rules out cubital tunnel syndrome. The best option would be ulnar demyelinating mononeuropathy that is not localizable with the tracing in Figure 14.3. The left ulnar forearm is less likely and left brachial plexopathy is less likely as most brachial plexopathies are axonal in nature. (**Preston and Shapiro 1998, p. 270**)

27. (A): Inching can be performed for ulnar nerve just like can be done in CTS. Any abrupt increase in latency or drop in amplitude between successive stimulation sites implies focal demyelination. In healthy individuals, the latency between two successive 1-cm stimulation sites does not change by >0.4 to 0.5 ms. Any greater latency shift suggests focal slowing. The inching technique is very sensitive but technically very demanding. The technique has the advantage of potentially being able to directly locate the lesion helping in deciding the best surgical technique to use. (**Preston and Shapiro 1998, p. 273**)

28. (C): The F-wave is a late motor response that occurs after the CMAP. It is derived by antidromic travel up the nerve to the anterior horn cell, with backfiring of a small population of anterior horn cell and orthodromic travel back down the nerve past the stimulation site to the muscle. Therefore F-wave is not a true reflex. If the stimulator is moved proximally the latency of F-wave response decreases. F-wave response varies slightly in latency, configuration, and amplitude because a different population of anterior horn cells is activated with each stimulation. Normal F-wave latency for ulnar nerve is 25 to 32 ms. (**Preston and Shapiro 1998, p. 45**)

29. (A): In this case, patient most likely had peroneal neuropathy from prolonged immobilization at the time of the surgery. The tracing shows a conduction block with drop in amplitude >20% across the fibular head and temporal dispersion suggesting focal demyelination at the fibular head. However, the peroneal distal motor latency is also abnormal and delayed suggesting underlying demyelinating possible peripheral neuropathy. The peroneal CMAP amplitude is abnormal and low suggesting secondary axonal involvement. (**Preston and Shapiro 1998, p. 310**)

30. (C): Midbrain lesion will not affect the blink reflex while pontine and medullary lesions will affect the afferent and the efferent arms of blink reflex.

Axonal neuropathies rarely affect the blink reflex because typical axonal distal dying-back neuropathies are unlikely to affect the blink reflex, which are so proximal. However in demyelinating neuropathies, all potentials of blink response may be markedly delayed or absent, reflecting slowing of either or both motor and sensory pathways. **(Preston and Shapiro 1998, p. 60)**

31. (B): In motor studies temporal dispersion and phase cancellation generally do not lead to appreciable drop in the proximal CMAP and area. In demyelinating lesions the conduction velocities may be very slow, and temporal dispersion and phase cancellation become more prominent for motor fibers. Using computer-simulation models, CMAP area has been demonstrated to fall by 50%, solely from the effects of temporal dispersion and phase cancellation in demyelinating lesions, without any conduction block. Now most electromyographers (EMGers) use the criteria of a >50% drop in amplitude to define electrophysiologic conduction block. **(Preston and Shapiro 1998, p. 42)**

32. (B): The EMG/nerve conduction studies (NCS) tracing showed a severely delayed median distal motor latency compared to ulnar distal motor latency. The ulnar distal motor latency is normal and there is no conduction block seen across the elbow. The comparison between median and ulnar distal motor latency is significant (>1.8 ms) and is therefore consistent with left CTS. Myasthenia gravis cannot be diagnosed with the tracings. Cervical radiculopathy can result is some delay in median distal motor latency but not to this extent especially without affecting the CMAP amplitude as most radiculopathies are axonal in nature. **(Preston and Shapiro 1998, p. 241)**

References

1. Preston DC, Shapiro BE. *Electromyography and neuromuscular disorders*, 1st ed. Butterworth-Heineman; 1998.

CHAPTER 15

Nerve Conduction Findings in Common Neuromuscular Disorders

QUESTIONS

1. **All statements about myasthenia gravis are correct except:**
   A. Approximately 50% of patients have the restricted ocular form of the disease
   B. If a patient's symptoms remain restricted to the ocular muscles for 2 years, then there is a high probability of it being restricted as ocular myasthenia
   C. Neonatal myasthenia is mild, self-limited, and disappears after the first few months of life
   D. Penicillamine can cause myasthenia, including the presence of acetylcholine receptor antibodies

2. **A decrement on repetitive nerve stimulation can be seen in following conditions except:**
   A. Neuropathies
   B. Motor neuron disease
   C. Congenital myopathies
   D. Myotonic disorder

3. **All statements regarding repetitive nerve stimulation (RNS) are normal except:**
   A. RNS are abnormal in 50% to 70% of patients with generalized myasthenia gravis
   B. RNS is usually abnormal in 60% of patients with ocular myasthenia gravis
   C. A compound motor action potential (CMAP) decrement of 10% or more is characteristic of myasthenia gravis
   D. Postexercise exhaustion is seen in patients with myasthenia gravis

4. **All statements regarding Lambert-Eaton myasthenic syndrome (LEMS) are correct except:**
   A. It is a rare condition
   B. It presents with diffuse muscle weakness
   C. CMAPs on routine motor nerve conduction studies are of large amplitude
   D. Deep tendon reflexes are characteristically reduced or absent

5. **All statements regarding RNS in LEMS are correct except:**
   A. Slow RNS (3 Hz) results in a decremental response similar to myasthenia gravis
   B. Rapid RNS (30 to 50 Hz) stimulation produces a marked increase in the CMAP amplitude
   C. Brief intense exercise produces a marked increase in the CMAP amplitude
   D. The incremental response seen in LEMS is 50% increase in the CMAP amplitude

6. **All statements about botulism are correct except:**
   A. Pathophysiology of botulism is presynaptic blocking of acetylcholine
   B. CMAP amplitudes are decreased with normal latencies and conduction velocities
   C. An incremental response occurs after brief exercise and fast RNS
   D. Lack of incrementing response to rapid RNS or brief exercise rules out the diagnosis of botulism

7. **Nerve conduction studies in myopathic disorder are:**
   A. Sensory conduction studies
   B. Motor nerve conduction studies
   C. F-wave
   D. Repetitive nerve stimulation

8. **Martin-Gruber anomaly is characterized by:**
   A. This is a very common anomaly, seen in two third of patients
   B. Most common anomaly shows decrease in ulnar CMAP amplitude (recording at abductor digiti minimi) with below-elbow stimulation compared to wrist stimulation

C. Most common anomaly shows decrease in ulnar CMAP amplitude (recording at first dorsal interosseous) with below-elbow stimulation compared to wrist stimulation
D. Most common anomaly shows decrease in increased median CMAP amplitude (recording at abductor policis brevis) with below-elbow stimulation compared to wrist stimulation

9. **Accessory peroneal nerve is characterized by:**
    A. Decrease in peroneal CMAP when stimulated at the fibular head compared to stimulation at the ankle
    B. Arises from deep peroneal nerve
    C. During routine nerve conduction studies, this anomaly is recognized during tibial motor studies
    D. The peroneal CMAP amplitude is lower when stimulated at the ankle compared to if stimulated behind the lateral malleolus

10. **Blink reflex is characterized by all except:**
    A. Both orbicularis oculi muscles are recorded simultaneously
    B. Each side the ipsilateral supraorbital nerve is stimulated over the medial eyebrow
    C. Only submaximal stimulation is used in getting blink reflex
    D. Ground electrode is placed on the chin

11. **An abnormal blink reflex is recorded with absent ipsilateral R1 and R2 potentials but a normal contralateral R2 potential on stimulating the affected side, whereas stimulating the unaffected side results in a normal ipsilateral R1 and R2 but absent contralateral R2. Where is the lesion?**
    A. Right medullary lesion
    B. Right midpontine lesion
    C. Right trigeminal lesion
    D. Right facial nerve lesion

12. **A patient presents with right hand numbness and tingling. The lesion is localized at (see Fig. 15.1):**
    A. Right carpal tunnel syndrome
    B. Artifact
    C. Right Martin-Gruber anomaly
    D. Right median neuropathy above the elbow

13. **A patient presents with left hand pain and tingling. The median motor and sensory conduction studies are normal. The ulnar nerve conduction study tracing is shown in Figure 15.2:**
    A. Left cubital tunnel syndrome
    B. Ulnar mononeuropathy, nonlocalizable
    C. Left ulnar mononeuropathy in the forearm
    D. Left brachial plexopathy

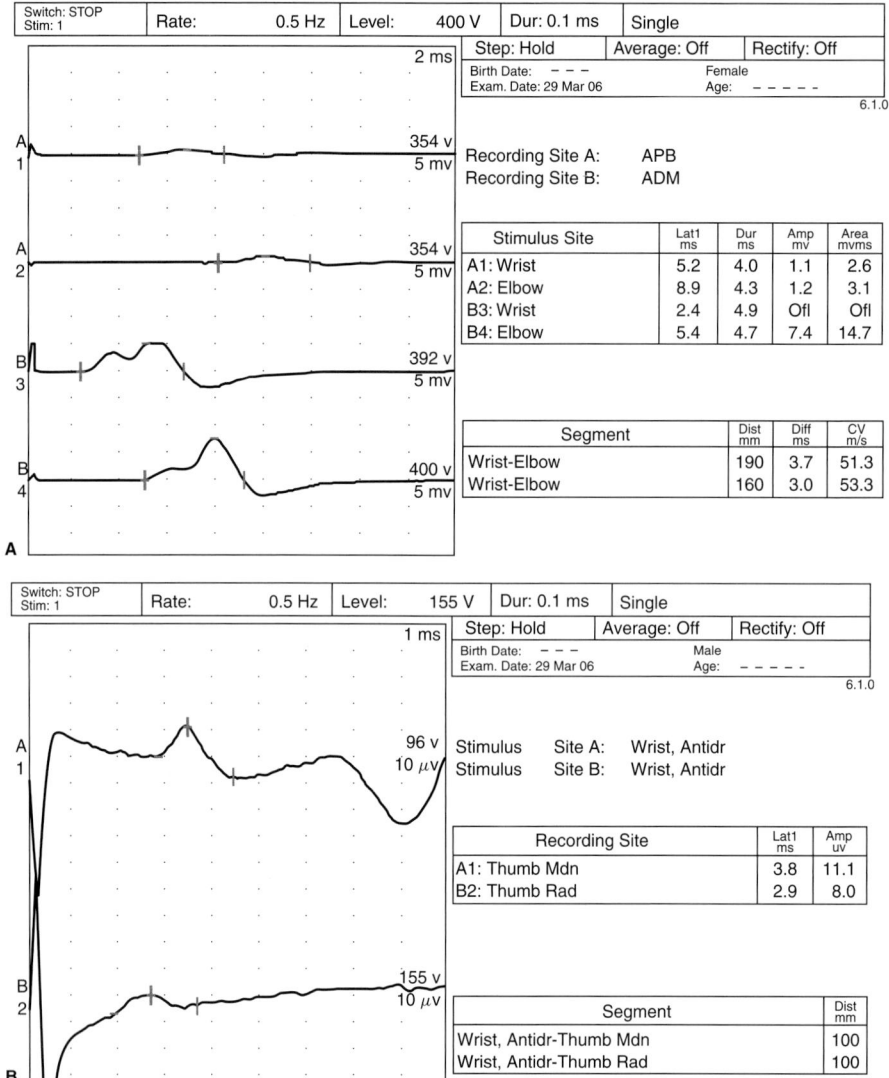

Figure 15.1. (**A**) Median versus radial sensory nerve conduction (SNC) and (**B**) median versus ulnar median nerve compartment (MNC). APB, abductor pollicis brevis; ADM, abductor digiti minimi

14. The above tracing in Figure 15.3 is from a patient with pain and numbness in the ulnar aspect of the hand. The inching techniques in ulnar nerve evaluation are preferred because:
   A. Any abrupt increase in latency or drop in amplitude between successive stimulation sites implies focal demyelination
   B. The latency difference of more than 0.2 ms is abnormal
   C. This is a very easy technique
   D. It does not add much except academic satisfaction

Chapter 15: Nerve Conduction Findings in Common Neuromuscular Disorders

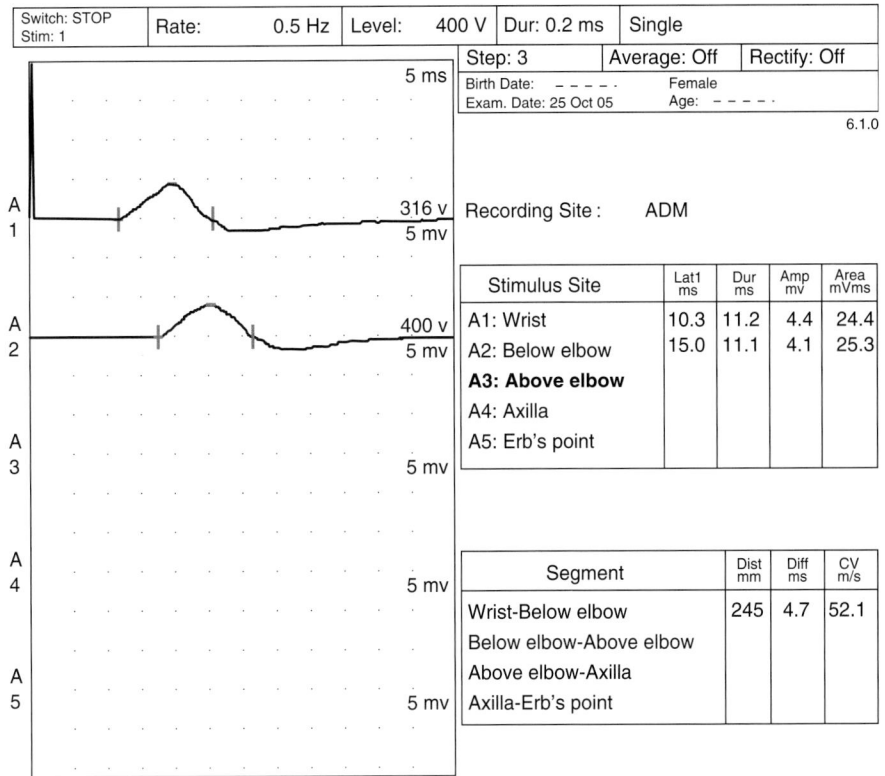

Figure 15.2. Ulnar median nerve compartment (MNC) demyelinating neuropathy

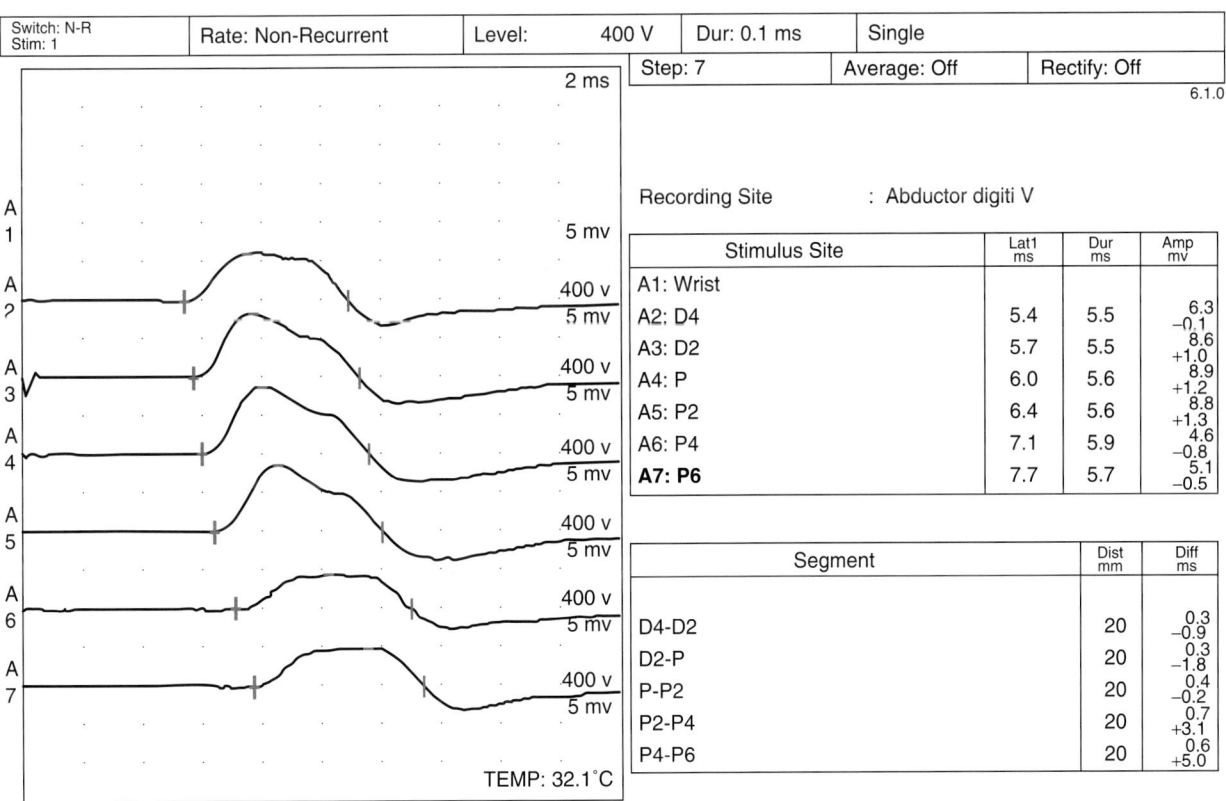

Figure 15.3. Ulnar inching

15. **The F-wave response in Figure 15.4 is normal. Choose the correct response related to F-waves:**
    A. If the stimulator is moved proximally, the latency of F-wave response increases
    B. The F-wave represents the reflex arc involving the sensory nerve dorsal horn, interneurons, anterior horn cell, and the motor nerve
    C. F-wave response varies slightly in latency, configuration, and amplitude because a different population of anterior horn cells is activated with each stimulation
    D. Normal F-wave latency for ulnar nerve is 40 ms

16. **Patient presents with right foot drop after having mitral valve replacement surgery. The right peroneal motor response is shown in Figure 15.5. All statements below are correct except:**
    A. Peroneal distal motor latency is normal
    B. There is significant conduction block seen across the fibular head
    C. The response at the fibular head shows temporal dispersion
    D. The CMAP amplitude is abnormal and low

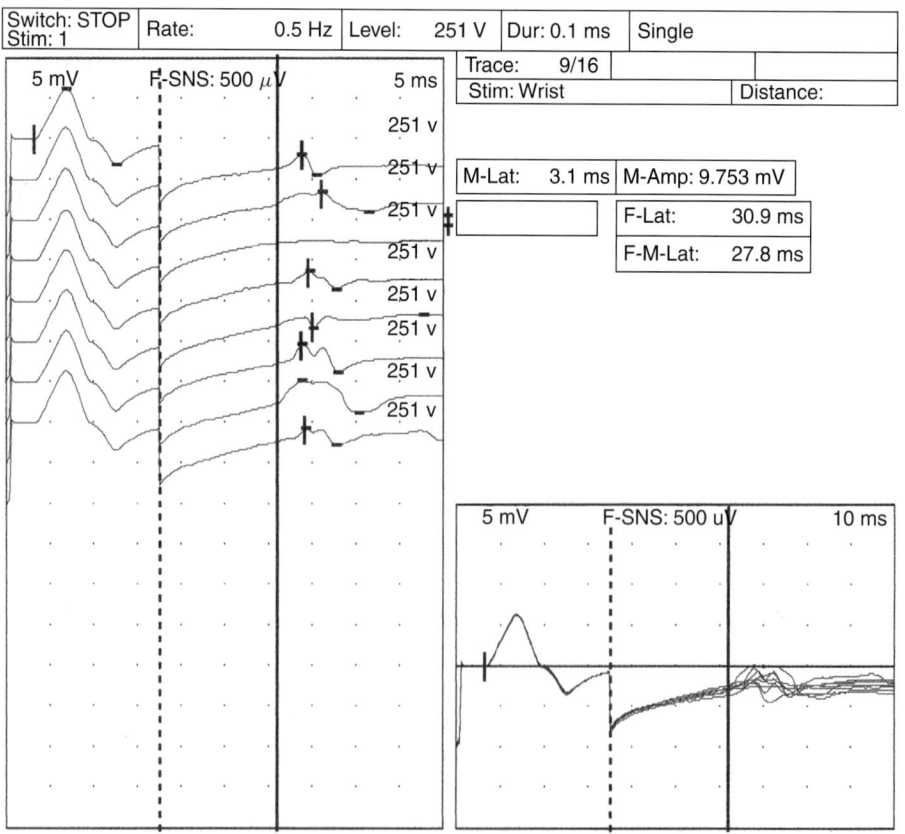

Figure 15.4. Ulnar F-wave. F-SNS, F-wave sensitivity

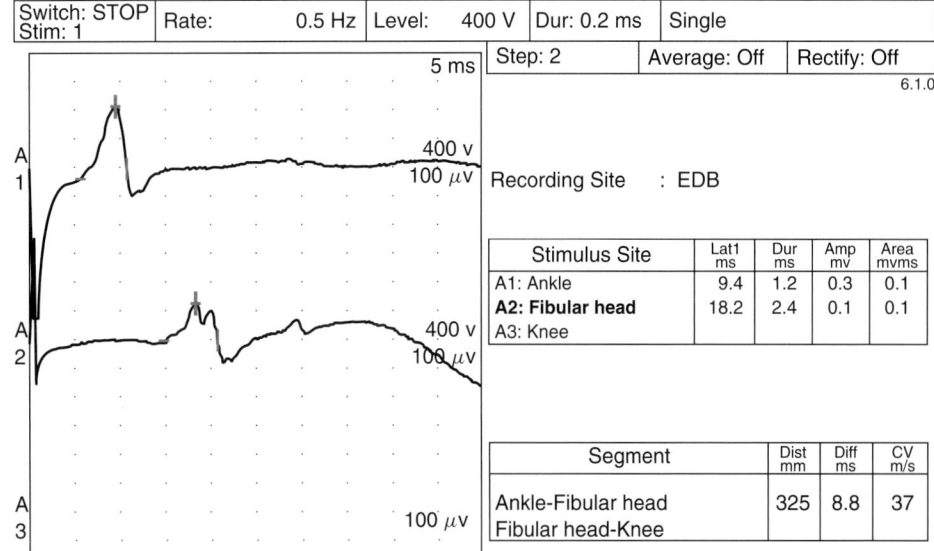

Figure 15.5. Peroneal block at fibular head. EDB, extensor digitorium brevis

17. Electrophysiologic studies in amyotrophic lateral sclerosis (ALS) are characterized by all except:
    A. Sensory nerve conduction studies are typically normal
    B. CMAP amplitude decreases as anterior horn cells die
    C. Motor nerve conduction velocity is normal
    D. Sensory nerve action potential (SNAP) amplitude may be mildly reduced

18. A decremental response with RNS can be seen with all except:
    A. Botulism
    B. Myotonic disorder
    C. McArdle disease
    D. Amyotrophic lateral sclerosis

19. Abnormal temporal dispersion with motor nerve conduction studies is consistent with:
    A. Neuromuscular junction disorder
    B. Demyelinating disorder
    C. Temperature effect
    D. Amplifier turned off

20. A 45-year-old female patient presented with lower back pain radiating in the right leg. You will evaluate her with nerve conduction study of all the following nerves except (see Fig. 15.6):
    A. Sural nerve
    B. Medial plantar nerve
    C. Peroneal nerve
    D. Tibial nerve

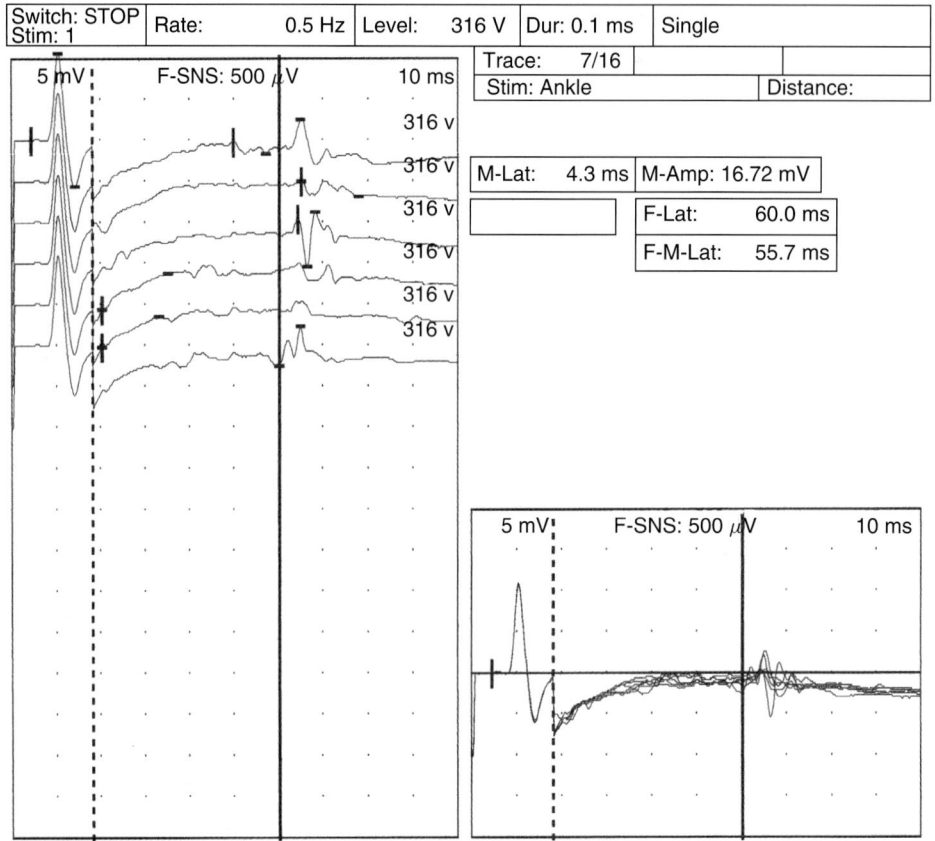

Figure 15.6. Tibial F-wave

21. A 33-year-old female presents with generalized weakness. We proceed with evaluation of sensory and motor nerve conduction studies. The results are as following:

    Sural, median, and ulnar sensory response—normal
    Right peroneal distal motor latency—5.4 ms, amplitude 4.9 mV
    Right tibial distal motor latency—4.8 ms, amplitude 1.5 mV
    Right median distal motor latency—3.5 ms, amplitude 18 mV
    Right ulnar distal motor latency—3.0 ms, amplitude 14 mV

    Your differential includes all of the following except:
    A. ALS
    B. Lambert-Eaton syndrome (LES)
    C. Myopathy
    D. Radiculopathy

22. The needle electromyography (EMG) of lower extremity, upper extremity, and paraspinal muscles is normal with normal motor units. You proceed with which one of the following tests to confirm your diagnosis:
    A. Thoracic paraspinal muscles
    B. Muscle biopsy
    C. Nerve biopsy
    D. Repetitive nerve stimulation

23. You do slow RNS test, which showed decremental response of CMAP. With brief intense exercise, the CMAP amplitude increased significantly. You proceed with single fiber EMG, which showed increased jitter and blocking. The final diagnosis is:
    A. ALS
    B. LES
    C. Myasthenia gravis
    D. Radiculopathy

24. A 21-year-old female presents with acute onset of left foot weakness and tingling paresthesia in both legs for 5 days. Her nerve conduction study showed:

    Sural, median, and ulnar sensory response—normal
    Right peroneal distal motor latency—5.4 ms, amplitude 4.9 mV
    Right tibial distal motor latency—4.8 ms, amplitude 8.5 mV
    Right median distal motor latency—3.5 ms, amplitude 18 mV
    Right ulnar distal motor latency—3.0 ms, amplitude 14 mV

    Right peroneal F-wave latency—60 ms
    Right tibial F-wave latency—59 ms
    Right median F-wave latency—29 ms
    Right ulnar F-wave latency—27 ms

    The needle EMG finding is normal. The best diagnosis is:
    A. Inflammatory myopathy
    B. ALS
    C. Guillain-Barré syndrome
    D. Myasthenia gravis

ANSWERS

1. **(A):** Only approximately 15% of patients have the restricted ocular form of the disease. If a patient's symptoms remain restricted to the ocular muscles for 2 years, then there is a high probability of it being restricted as ocular myasthenia. Neonatal myasthenia develops when maternal autoantibodies cross the placenta and result in clinical syndrome in a newborn infant. Neonatal myasthenia is mild, self-limited, and disappears after the first few months of life. Penicillamine can cause myasthenia, including the presence of acetylcholine receptor antibodies. Myasthenia gravis may also be seen in patients treated with penicillamine. The clinical syndrome is similar to myasthenia gravis, including the presence of acetylcholine receptor antibodies except that most patients slowly improve with penicillamine discontinuation. **(Preston and Shapiro 1998, p. 506)**

2. **(C):** In any patient suspected of myasthenia gravis, routine motor and sensory nerve conductions must be done. Particular attention must be paid to normal CMAP amplitude in myasthenia in comparison to LES, where baseline CMAP is usually diffusely low. A decrement on repetitive nerve stimulation can also be seen in various conditions, for example, neuropathies, motor neuron disease,

inflammatory myopathies, and myotonic disorders. A decrement on repetitive nerve stimulation of the ulnar nerve may be seen in a severe ulnar neuropathy with denervation; such a finding in the context does not imply a primary neuromuscular junction disorder. (**Preston and Shapiro 1998, p. 506**)

3. (**B**): RNS is typically abnormal in 50% to 70% of patients with generalized myasthenia gravis. In most patients with ocular myasthenia RNS is normal. In normal subjects, slow RNS (3 Hz) results in little or no decrement of CMAP, whereas in myasthenia gravis a CMAP decrement of 10% or more is characteristically seen. The diagnostic yield increases with proximal nerves. If baseline RNS in myasthenia gravis is inconclusive, a postexercise after 1 minute shows significant decrement in the CMAP. (**Preston and Shapiro 1998, p. 506**)

4. (**C**): LEMS is quite rare. It affects males more than females usually older than 40 years. It presents with generalized weakness more in the proximal muscles. The deep tendon reflexes are characteristically reduced or absent, which is unusual in myasthenia gravis.

Single stimuli produce a reduced release of acetylcholine and at rest many of the end-plate potentials do not reach threshold, resulting in small-amplitude CMAPs on routine motor nerve conduction studies. (**Preston and Shapiro 1998, p. 510**)

5. (**D**): The electrophysiology of LEMS is diagnostic. Rapid RNS (30- to 50-Hz) or brief intense exercise produces a marked increase in the CMAP amplitude due to calcium accumulation in the presynaptic nerve terminal with subsequent enhancement of release of acetylcholine. The CMAP commonly increases in amplitude by >200%. Brief intense exercise is preferred over rapid RNS because rapid RNS can be painful and is not easily tolerated by the patients. This marked postexercise facilitation of CMAP is the electrical correlate of the clinical facilitation of muscle strength and reflexes seen after brief exercise. (**Preston and Shapiro 1998, p. 510**)

6. (**D**): The pathophysiology of botulism is presynaptic blocking of acetylcholine.

Like Lambert-Eaton myasthenic syndrome, CMAP amplitudes are decreased with normal latencies and conduction velocities. A decremental response is seen with slow RNS (3 Hz). An incremental response occurs after brief exercise and fast RNS in early or mild cases. However in severe botulism, if amount of acetylcholine release has dropped severely below threshold, even facilitation with rapid RNS or brief exercise may not result in a threshold response and therefore no increment occurs in the CMAP amplitude. Lack of incrementing response to rapid RNS or brief exercise cannot rule out the diagnosis of botulism. (**Preston and Shapiro 1998, p. 512**)

7. (**B**): Routine nerve conduction studies should always be done in patients with suspected myopathy. Sensory conduction studies are always normal unless there is a coexistent neuropathy. Motor conduction studies are usually normal as myopathies usually involve proximal muscles that are not tested in routine motor nerve conduction studies. However, if the myopathy is severe enough to affect distal and proximal muscles or one of the rare myopathies that preferentially

affects distal muscles, motor nerve studies may be *abnormal* and *show decreased CMAP* with normal latencies and conduction velocities. (**Preston and Shapiro 1998, p. 527**)

8. **(C):** The most commonly encountered anomaly in the upper extremity is a crossover of median-to-ulnar fibers, the Martin-Gruber anastomosis (MGA) seen in approximately 15% to 30% of patients. There are three types of MGA anomalies seen: (a) to innervate hypothenar muscles (shows decrease ulnar CMAP amplitude [recording at abductor digiti minimi] with below-elbow stimulation compared to wrist stimulation), (b) to innervate the first dorsal interosseous muscle (this is the most common MGA anomaly and shows decrease in ulnar CMAP amplitude [recording at first dorsal interosseous] with below-elbow stimulation compared to wrist stimulation, and (c) to innervate thenar muscles (shows decrease in increased median CMAP amplitude [recording at abductor policis brevis] with below-elbow stimulation compared to wrist stimulation). (**Preston and Shapiro 1998, p. 78**)

9. **(D):** The most common anomalous innervation in the leg is an accessory peroneal nerve in the lateral calf. The extensor digitorum brevis (EDB) muscle gets extra innervation by an anomalous motor branch originating from the superficial peroneal nerve. During the routine nerve conduction studies, this anomaly is recognized during peroneal motor studies. The accessory peroneal nerve travels down the lateral calf, posterior to the lateral malleolus. If anomaly is present, the peroneal CMAP amplitude recording the EDB is higher stimulating below the fibular head and lateral malleolus than at the ankle. (**Preston and Shapiro 1998, p. 81**)

10. **(C):** The blink reflex study measures the entire reflex arc between the trigeminal and facial nerves, including proximal segment of facial nerve. To get the response, both orbicularis oculi muscles are recorded simultaneously. On each side the ipsilateral supraorbital nerve is stimulated over the medial eyebrow. It is crucial to record 4 to 6 supramaximal stimulation trials. The ground electrode is placed on the chin. It is extremely important that the patient be in a relaxed state to eliminate any signal noise.
The R1 latency represents conduction time along the fastest fibers of the afferent pathway of ipsilateral trigeminal nerve and R2 latency represents conduction time from ipsilateral trigeminal nucleus to both ipsilateral and contralateral facial nerve nuclei and in turn to facial nerves. Normal R1 and R2 latencies are 10 to 12 ms and 30 to 40 ms, respectively. (**Preston and Shapiro 1998, p. 58**)

11. **(D):** For each blink response, the absolute R1 and R2 latencies are compared with normal controls as well as with the contralateral side. Normal R1 and R2 latencies are 10 to 12 ms and 30 to 40 ms, respectively. Stimulating the affected side results in absence of ipsilateral R1 and R2 potentials but a normal contralateral R2 potential, whereas stimulating the unaffected side results in a normal ipsilateral R1 and R2 but absent contralateral R2, suggests a unilateral facial nerve lesion. In this pattern all potentials on the affected side are abnormal, regardless of which side is stimulated. In demyelinating peripheral polyneuropathy, all potentials of the blink response may be markedly delayed or

absent, reflecting slowing of either or both motor and sensory pathways. (**Preston and Shapiro 1998, p. 60**)

12. (A): The tracings in Figure 15.1A showed a delayed median antidromic sensory latency compared to radial sensory latency. Figure 15.1B showed a prolonged median distal motor latency and absence of conduction block. These findings are consistent with a right carpal tunnel syndrome. MGA is the most commonly encountered anomaly in the upper extremity with a crossover of median-to-ulnar fibers. The MGA is seen in approximately 15% to 30% of patients. The most common MGA anomaly shows decrease in ulnar CMAP amplitude (recording at first dorsal interosseous) with below-elbow stimulation compared to wrist stimulation. We do not see this in the figure hence it is not a correct option. (**Preston and Shapiro 1998, p. 78**)

13. (B): This above tracing in Figure 15.2 shows ulnar motor response at wrist and at elbow. The ulnar distal motor latency is severely delayed with decreased amplitude. This is consistent with ulnar demyelinating ulnar motor neuropathy with secondary axonal involvement (indicated by low amplitude). There is no conduction block seen across the elbow. This effectively rules out cubital tunnel syndrome. The best option would be an ulnar demyelinating mononeuropathy that is not localizable with the above tracing. The left ulnar forearm is less likely and left brachial plexopathy is less likely because most brachial plexopathies are axonal in nature. (**Preston and Shapiro 1998, p. 273**)

14. (A): Inching can be performed for ulnar nerve just as that done in carpal tunnel syndrome. Any abrupt increase in latency or drop in amplitude between successive stimulation sites implies focal demyelination. In normal individuals, the latency between two successive 1-cm stimulation sites does not change by 0.4 to 0.5 ms. Any greater latency shift suggests focal slowing. The inching is very sensitive but technically very demanding. The technique has the advantage of potentially being able to directly locate the lesion helping in deciding the best surgical technique to use. (**Preston and Shapiro 1998, p. 273**)

15. (C): The F-wave is a late motor response that occurs after the CMAP. It is derived by antidromic travel up the nerve to the anterior horn cell, with backfiring of a small population of anterior horn cell and orthodromic travel back down the nerve past the stimulation site to the muscle. Therefore, F-wave is not a true reflex. If the stimulator is moved proximally, the latency of F-wave response decreases. F-wave response varies slightly in latency, configuration, and amplitude because a different population of anterior horn cells is activated with each stimulation. Normal F-wave latency for ulnar nerve is 25 to 32 ms. (**Preston and Shapiro 1998, p. 45**)

16. (A): In this case, patient most likely had peroneal neuropathy from prolonged immobilization at the time of the surgery. The tracing shows a conduction block with a drop in amplitude >20% across the fibular head and temporal dispersion suggesting focal demyelination at the fibular head. However, the peroneal distal motor latency is also abnormal and delayed suggesting underlying demyelinating possible peripheral neuropathy. The peroneal CMAP amplitude is abnormal and low suggesting secondary axonal involvement. (**Preston and Shapiro 1998, p. 310**)

17. (C): Electrophysiologic studies are considered essential for evaluation of a patient suspected for ALS. Sensory nerve conduction studies are typically normal; however, some studies have shown mildly reduced SNAP amplitude and sensory nerve conduction velocity. CMAP amplitude decreases as anterior horn cells die and muscle fibers lose their innervation. As the CMAP amplitude becomes severely reduced, motor nerve conduction velocity often slows due to loss of the large myelinated fibers but not to <80% of the lower limit of normal. (**Ross 2002, p. 25**)

18. (A): A decremental response with RNS is seen predominantly in primary disorders of the neuromuscular junction. However, a decrement may be seen in other disorders, especially in severe denervating disorders, for example, ALS. In any condition in which there is prominent denervation and reinnervation, newly formed neuromuscular junctions (NMJs), which occur as denervated fibers are reinnervated, are immature and unstable. These immature and unstable NMJs may show a decremental response with RNS. Some myopathic conditions including myotonic disorders and metabolic myopathies like McArdle disease may also show a decremental response. (**Preston and Shapiro 1998, p. 72**)

19. (B): In motor studies, temporal dispersion and phase cancellation generally do not lead to appreciable drop in the proximal CMAP area. In demyelinating lesions, the conduction velocities may be very slow, and temporal dispersion and phase cancellation become more prominent for motor fibers. Using computer-simulation models, CMAP area has been demonstrated to fall by 50%, solely from the effects of temporal dispersion and phase cancellation in demyelinating lesions, without any conduction block. Now most EMGers use the criteria of a >50% drop in amplitude to define electrophysiologic conduction block. (**Preston and Shapiro 1998, p. 42**)

20. (B): All the above nerves will help to see if there is radiculopathy while medial plantar nerve will not add much information that is not already obtained by the above nerves. Her sural sensory latency was normal. Peroneal distal motor latency was 5.5 ms with normal amplitude. Tibial distal motor latency was 5.8 ms with normal amplitude with borderline normal F-wave latency (Fig. 15.6). (**Preston and Shapiro 1998, p. 417**)

21. (C): The CMAP amplitude is decreased in peroneal and tibial nerve. Severe inflammatory myopathy can result in some drop in CMAP amplitude but not to this degree and therefore myopathy is excluded from the differential. All the other options should be evaluated in a generalized neuropathic process. (**Preston and Shapiro 1998, p. 526**)

22. (D): Because we have ruled out motor neuron disease and radiculopathy, next consideration would be of a neuromuscular junction disorder. Repetitive nerve stimulation abnormality is the hallmark of neuromuscular junction disorder. (**Preston and Shapiro 1998, p. 506**)

23. (B): In LEMS, slow RNS results in a decremental response similar to myasthenia gravis. The electrophysiology of LEMS is diagnostic. Rapid RNS (30- to 50-Hz) or brief intense exercise produces a marked increase in the CMAP amplitude due to calcium accumulation in the presynaptic nerve terminal with subsequent enhancement of release of acetylcholine. The CMAP commonly increases in amplitude by >200%. Brief intense exercise is preferred over rapid RNS, because rapid RNS can be painful and is not easily tolerated by the patients. This marked postexercise facilitation of CMAP is the electrical correlate of the clinical facilitation of muscle strength and reflexes seen after brief exercise. **(Preston and Shapiro 1998, p. 510)**

24. (C): The distal motor latencies are normal while the peroneal and tibial F-wave latency are slightly prolonged. The median and ulnar distal motor latency was normal. This pattern is very consistent with Guillain-Barré syndrome. In the first week of the disease, usually the nerve conduction study and the EMG study is normal. However, the only abnormality seen is slightly delayed F-wave latencies in the lower extremity. However, as time passes Wallerian degeneration occurs and acute denervation pattern in seen in EMG. It is advisable to repeat the study in 2 to 3 weeks after the first study. **(Preston and Shapiro 1998, p. 363)**

References

1. Preston DC, Shapiro BE. *Electromyography and neuromuscular disorders*, 1st ed. Butterworth-Heineman; 1998.
2. Ross MA. *Continuum.* 2008; 8(4).

SECTION 5

Electromyography

CHAPTER 16

General Principles of Electromyography

QUESTIONS

1. A concentric electromyographic (EMG) needle has the following benefits over a monopolar needle except:
   A. The major spike rise time is shorter than those obtained with a monopolar needle
   B. It does not require an additional reference electrode
   C. It is easier for patients to tolerate
   D. Motor unit action potential (MUAP) amplitude is slightly smaller

2. Needle movement resulting in any abnormal waveform is called *increased insertional activity* if it lasts longer than:
   A. 3 ms
   B. 30 ms
   C. 300 ms
   D. 3,000 ms

3. The end-plate noise:
   A. Is an abnormal spontaneous activity
   B. Is of high amplitude
   C. Has monophasic positive potentials
   D. Has a characteristic "seashell" sound

175

4. **All statements regarding fibrillation potentials are correct except:**
   A. Spontaneous activity of a single muscle fiber
   B. Spontaneous depolarization of muscle fibers
   C. Neuropathic process
   D. Ocular myasthenia gravis

5. **All statements regarding fibrillation potentials are correct except:**
   A. Are recognized by brief spikes with initial positive deflection
   B. Are very regular in firing pattern
   C. Increase after moderate cooling of the muscle
   D. Are of low amplitude

6. **All statements regarding positive waves are correct except:**
   A. Spontaneous depolarization of muscle fibers
   B. Signify denervation
   C. Of positive polarity followed by a long negative phase
   D. Of an irregular firing pattern

7. **All statements regarding complex repetitive discharges (CRDs) are correct except:**
   A. Result from the depolarization of a single muscle fiber followed by ephaptic spread to adjacent denervated fibers.
   B. Usually occur spontaneously and are of high frequency
   C. Are repetitive discharges with an abrupt onset and termination
   D. Are variable in morphology from one discharge to the next

8. **All statements regarding myotonic discharges are correct except:**
   A. Spontaneous discharge of a muscle fiber
   B. Characterized by waxing and waning of both amplitude and frequency
   C. Commonly associated with CRDs
   D. May occur in some myopathies

9. **Fasciculations:**
   A. Represent single, spontaneous, involuntary discharge of an individual motor unit
   B. Generally fire very fast and regularly
   C. May have morphology of simple or complex MUAP
   D. The source generator is motor neuron or its axon

10. **Which of the following statements is not true about the differentiation of benign fasciculations from malignant fasciculations?**
    A. Benign fasciculations are not associated with muscle weakness
    B. Benign fasciculations are not associated with wasting
    C. Benign fasciculations affect the same site repetitively
    D. Benign fasciculations tend to fire slower

11. **All statements regarding myokymic discharges are correct except:**
    A. They are spontaneous repetitive discharges of the same motor unit
    B. The firing frequency between bursts is slower

C. Changing to a shorter sweep helps to recognize the myokymic discharge
D. They produce a marching sound on EMG

12. **The following statements are correct except:**
    A. EMG of cramps show full interference pattern of MUAP with normal morphology
    B. EMG of contractures show full interference pattern of MUAP
    C. Neuromyotonic discharges are higher frequency, decrementing repetitive discharge of a single motor unit
    D. Neuromyotonic discharges persist in sleep

13. **Acute myopathies of less than 3 weeks in duration on EMG are characterized by:**
    A. Polyphasic large motor units
    B. Polyphasic shorter duration and smaller amplitude motor units
    C. Delayed recruitment with rapid firing
    D. Positive waves and fibrillations

14. **The key factor that differentiates nascent motor units from myopathic motor units is:**
    A. Amplitude
    B. Morphology
    C. Recruitment pattern
    D. Sound of EMG

15. **EMG of tremor is characterized by:**
    A. Same MUAP fires repetitively and continuously
    B. Bursting pattern of involuntary MUAP
    C. Different MUAPs fire repetitively and continuously
    D. MUAP cannot be suppressed by positioning

16. **The rate of axonal regrowth is:**
    A. Similar to rate of demyelinating regrowth
    B. Similar for distal upper and lower extremities
    C. Is approximately 1 cm per day
    D. Is approximately 1 mm per day

17. **An acute axonal injury is defined as:**
    A. Injury within a few hours
    B. Injury within a few days
    C. Injury within a few weeks
    D. Injury within a few months

18. **Spontaneous activity on needle EMG examination is:**
    A. Always abnormal
    B. Normal only if potential is recorded furthest from the motor end plate
    C. Normal only if recorded in proximity to the muscle end plate
    D. None of the above

19. **The characteristic sound of end-plate noise potential is:**
    A. Buzzing sound
    B. Rain on the roof
    C. Cracking sound
    D. Seashell sound

20. **End-plate spikes are differentiated from fibrillation potentials by all except:**
    A. Initial positive deflection
    B. 50 Hz frequency
    C. Irregular firing rate
    D. Association with end-plate noise

21. **Fibrillation potentials are characterized by all of the following except:**
    A. Extracellular recording of a single motor fiber
    B. They are only observed in neurogenic diseases
    C. Rain on the roof sound
    D. Regular firing pattern with an initial positive deflection

22. **Positive waves are characterized by:**
    A. Regular firing pattern
    B. Extremely short duration
    C. A hissing sound
    D. Waxing and waning of frequency

23. **Complex repetitive discharges are characterized by:**
    A. Low frequency discharge
    B. Waxing and waning in frequency and amplitude
    C. Persistence with neuromuscular junction blockage
    D. Occurrence solely in myopathic diseases

24. **A spontaneous discharge of a single muscle fiber with a waxing and waning frequency and amplitude is a:**
    A. Fasciculation
    B. Myotonic discharge
    C. Myokymic discharge
    D. Complex repetitive discharge

25. **Fasciculations are characterized by all of the following except:**
    A. Spontaneous involuntary discharge of an individual motor fiber
    B. Always abnormal
    C. Grouped fasciculations are rhythmic, called myokymic discharges
    D. Commonly seen in radiculopathies, polyneuropathies, and entrapment neuropathies

26. **Myokymic responses are characterized by all of the following except:**
    A. Are rhythmic group repetitive discharges of the same motor unit
    B. May be a sign of radiation plexopathy in cancer patients
    C. Occur in 15% of patients with acute Guillain-Barré syndrome as facial myokymia
    D. Are inhibited with hypocalcemia

27. **All statements regarding monopolar needles are true except:**
    A. Do not have a beveled tip
    B. The shaft of the needle is the active electrode
    C. Smaller caliber than concentric needle
    D. Aluminum wire

28. **All statements about single fiber EMG are correct except:**
    A. A normal single fiber examination of a clinically weak muscle effectively rules out the diagnosis of myasthenia gravis
    B. The low frequency filter should be 10 Hz for recording single fiber EMG
    C. Single fiber needle is a modified concentric needle electrode
    D. The single fiber muscle action potentials are only recorded if the muscle action potentials are within 200 to 300 μm

29. **Which of the following forms of spontaneous activity in needle EMG is normal?**
    A. Myotonia
    B. Positive sharp waves
    C. Fibrillation potentials
    D. End-plate potentials

ANSWERS

1. **(C):** The MUAP amplitude is slightly smaller than what is obtained with a monopolar needle. However the major spike rise time is shorter with a concentric needle than those obtained with a monopolar needle. The concentric needle does not require an additional reference electrode and is easier for patients to tolerate. However the thin Teflon-coated monopolar needles are better tolerated than the concentric needles. (**Preston and Shapiro 1998, p. 146; Jun Kimura, p. 39**)

2. **(C):** Insertion of a needle electrode into the muscle normally gives rise to brief bursts of electrical activity. The same discharges also occur with each repositioning. It appears as positive or negative high-frequency spikes in a cluster. Increased insertional activity may be seen in both neuropathic and myopathic conditions. In case of muscle being replaced by fat and fibrous tissue, insertional activity may actually be decreased. (**Preston and Shapiro 1998, p. 182; Jun Kimura, p. 39**)

3. **(D):** End-plate noise is low-amplitude, monophasic negative potentials that fire irregularly at 20 to 40 Hz and have a characteristic seashell sound on EMG. It represents extracellularly recorded miniature end-plate potentials (MEPPs), nonpropagating depolarization caused by spontaneous release of acetylcholine quanta. (**Preston and Shapiro 1998, p. 183; Jun Kimura, p. 231**)

4. **(D):** Fibrillation potential is derived from the extracellular recording of a single muscle fiber. These spontaneous depolarizations of muscle fibers are markers of denervation. They are typically associated with neurogenic disorders, may also be

seen in muscle diseases and rarely in severe neuromuscular junctions disorders unless the condition is poorly controlled (especially in botulism). (**Preston and Shapiro 1998, p. 184; Jun Kimura, p. 258**)

5. (C): Fibrillations are brief spikes with initial positive deflection. They are of 1 to 5 ms in duration and low amplitude. Their firing pattern is very regular (0.5 to 10 Hz). The fibrillation potentials increase after warming the muscle or with administration of cholinesterase inhibitors. During cooling of the limb or with hypoxia the fibrillation discharges decrease. (**Preston and Shapiro 1998, p. 184; Jun Kimura, p. 258**)

6. (D): Positive waves are spontaneous depolarization of muscle fiber and signify denervation. They have positive polarity followed by a long negative phase. The absence of a negative spike implies recording near the damaged part of the muscle fiber. Positive waves are usually seen with fibrillation potentials but may be seen alone, sometimes early in denervation. The positive waves have variable amplitude but have a regular firing pattern. (**Preston and Shapiro 1998, p. 184; Jun Kimura, p. 259**)

7. (D): CRDs result from spontaneous depolarization of a group of muscle fibers firing in near synchrony. They result from the depolarization of a single muscle fiber followed by ephaptic spread to adjacent denervated fibers. The entire sequence repeats itself at slow or fast rates. The polyphasic and complex waveform remains uniform from one discharge to another. They have an abrupt onset and termination. CRDs less commonly are triggered by a stimulated MUAP or by a voluntary MUAP. (**Preston and Shapiro 1998, p. 186; Jun Kimura, p. 262**)

8. (C): A myotonic discharge is a spontaneous discharge of a muscle fiber and is characterized by its waxing and waning of amplitude and frequency. The firing rate is generally between 20 to 150 Hz. An individual myotonic potential may have a positive wave or brief spike morphology. Myotonic discharges are characteristically seen in myotonic dystrophy, myotonia congenita, and paramyotonia congenita. They may occur in some myopathies and in hyperkalemic periodic paralysis. (**Preston and Shapiro 1998, p. 186; Jun Kimura, p. 253**)

9. (B): A fasciculation is a single, spontaneous, involuntary discharge of an individual motor unit. It generally fires very slowly and irregularly between 0.1 to 10 Hz. The source generator is the motor neuron or its axon, proximal to its terminal. Fasciculations have the morphology varying from a simple MUAP to a complex and large if they represent a pathologic motor unit. Actual site of origin of most fasciculations has been found to be distal axon. They are brief twitches that seldom result in significant movement of a joint. (**Preston and Shapiro 1998, p. 187**)

10. (D): Distinguishing "benign" from "malignant" fasciculations on a clinical basis is nearly impossible. However, benign fasciculations are not associated with muscle weakness, wasting, or reflex changes. Benign fasciculations tend to fire faster and to affect the same site repetitively as opposed to malignant fasciculations that are more random. (**Preston and Shapiro 1998, p. 29; Jun Kimura, p. G626**)

11. **(C):** Myokymic discharges are rhythmic, grouped, spontaneous repetitive discharges of the same motor unit. The firing frequency within the burst is typically 5 to 60 Hz. The number of potential within a burst varies widely and may change from burst to burst. In between the bursts, the firing frequency is much slower. The myokymic discharges produce a marching sound on EMG. Changing to a longer not shorter sweep helps to recognize the myokymic discharge. **(Preston and Shapiro 1998, p. 188; Jun Kimura, p. 259)**

12. **(B):** Neuromyotonic discharges are bursts of motor unit action potentials, which originate in the motor axons firing at high rates for a few seconds and which often start and stop abruptly. The neuromyotonic discharges persist in sleep and typically have high frequency and a decrementing and waning pattern. EMG of cramps show full interference pattern of MUAP with normal morphology. This EMG pattern differentiates cramps from contractures. The EMG of contractures typically shows complete electrical silence. **(Preston and Shapiro 1998, p. 189)**

13. **(B):** In acute myopathies, the number of functioning muscle fibers in a motor unit decreases. This results in polyphasic shorter duration and smaller amplitude motor units. The other characteristic is early recruitment pattern. Polyphasic large motor units and delayed recruitment with rapid firing is a feature of neuropathic processes such as neuropathies, anterior horn cell diseases, and radiculopathies. In chronic myopathies, especially those with necrotic or inflammatory features, positive waves and fibrillations are commonly seen. **(Preston and Shapiro 1998, p. 202)**

14. **(C):** Early reinnervated motor units following severe denervation are known as *nascent motor units*. The key factor that differentiates nascent motor units from myopathic motor units is the recruitment pattern. Nascent MUAPs are always seen in the context of markedly reduced recruitment, whereas myopathic MUAPs are seen in the context of normal or early recruitment. **(Preston and Shapiro 1998, p. 203)**

15. **(C):** Tremor is recognized by a burst pattern of multiple MUAP firing simultaneously. The morphology of individual MUAP is difficult to assess as they tend to clump together. When tremor happens at rest, it can be mistaken for myokymia. In myokymia the same MUAP fires repetitively and continuously, whereas in tremor different MUAPs fire. In addition most patients can voluntarily alter their tremor by changing their limb position, whereas myokymia cannot be influenced by the patient. **(Preston and Shapiro 1998, p. 205)**

16. **(D):** The rate of axonal regrowth is limited by slow axonal transport. It is approximately 1 mm per day, so upper extremity axonal lesions will have a faster rate of recovery compared to lower extremities. It is much slower than the remyelination rate which only takes few weeks. Therefore, a demyelinating lesion with secondary axonal loss will have a more guarded prognosis than a purely demyelinating lesion. **(Kimura, p. 68)**

17. **(C):** Hyperacute injury is defined as injury within one week, acute injury is within few weeks, subacute injury is weeks to few months and chronic injury is defined as injury for more than few months. Enough time is passed for Wallerian

degeneration to have occurred. Accordingly, nerve conduction studies are abnormal, amplitudes are decreased with relatively normal conduction velocities and distal latencies, unless some of the largest and fastest axons have been lost, in which case some slowing of velocity and latency occurs. (**Preston and Shapiro 1998, p. 213**)

18. (C): Spontaneous activity (defined as activity lasting for >3 seconds) is abnormal except for potentials that occur in the muscle end-plate region (i.e., the neuromuscular junction [NMJ]). It is found near the center of the muscle belly and is often accompanied by a deep burning sensation. (**Preston and Shapiro 1998, p. 182**)

19. (D): End-plate noise represents low-amplitude monophasic negative potentials with a characteristic seashell sound on needle EMG. They have a firing rate of 20 to 40 Hz and physiologically represent MEPPs. (**Preston and Shapiro 1998, p. 183**)

20. (A): End-plate spikes are caused by irritation of the terminal twigs by the needle causing the propagation of an action potential. Since they are recorded at the site of stimulation of the action potential, they have a characteristic initial negative deflection and have a highly irregular pattern. (**Preston and Shapiro 1998, p. 183**)

21. (B): Fibrillation potentials are recognized by their regular firing rate (0.5 to 10 Hz) and their morphology of a single motor fiber with an initial positive deflection. They are markers of neurogenic diseases (neuropathies, radiculopathies, motor neuron disease, etc.) but can also occur in myopathies and in severe NMJ diseases like botulism. When chronic, they become small (<10 μV). (**Preston and Shapiro 1998, p. 184**)

22. (A): Positive waves are similar to fibrillation potentials as they also represent spontaneous depolarization of single muscle fiber. They are also regular and have a small initial positive deflection followed by a prolonged negative deflection, which is responsible for the dull pop sound. (**Preston and Shapiro 1998, p. 184**)

23. (C): CRDs occur by a spontaneous depolarization of a single muscle fiber followed by emphatic spread to adjacent denervated muscle fibers. It has a characteristic "machine like sound" of 20 to 150 Hz in frequency. It occurs in both chronic myopathic and neurogenic diseases. CRDs persist after neuromuscular junction blockage. (**Preston and Shapiro 1998, p. 185**)

24. (B): A myotonic discharge is a spontaneous discharge of a muscle fiber similar to fibrillations and positive waves but different in that it waxes and wanes in frequency and amplitude. It might have the morphology of a positive wave or fibrillation. They are characteristically seen in myotonic dystrophy, myotonia congenita, and paramyotonia congenita. They also occur in certain myopathies like acid maltase deficiency and hyperkalemic periodic paralysis. (**Preston and Shapiro 1998, p. 186**)

25. (B): Fasciculations are spontaneous involuntary discharge of an individual motor unit. They have an irregular and slow firing pattern with a single motor unit morphology. They are mostly seen in motor neuron diseases like amyotrophic lateral sclerosis (ALS), radiculopathies, polyneuropathies, and entrapment neuropathies. They are often normal in the absence of weakness and may persistence in one region (e.g., eyelid twitching). (**Preston and Shapiro 1998, p. 187**)

26. (**D**): Myokymia from peripheral nerve lesions can be provoked or enhanced by lowering of the serum-ionized calcium manifested as tetany. They occur with hyperventilation or use of acid-citrase-dextrose (ACD) anticoagulant as is commonly given during plasma exchange. Hyperventilation-induced myokyma manifests as carpopedal spasm. (**Preston and Shapiro 1998, p. 188; Jun Kimura, p. 259**)

27. (**B**): The MUAP amplitude is slightly smaller with a concentric needle than what is obtained with monopolar needle. However, the major spike rise time is shorter with concentric needle than those obtained with a monopolar needle. The concentric needle does not require an additional reference electrode and is easier for patients to tolerate. The monopolar needle is Teflon coated and its exposed tip serves as the active electrode. The Teflon coating can peel off, exposing the shaft, resulting in decrease in amplitude of MUAP without any significant effect on its duration. The major disadvantage of the monopolar needle is the need for an additional reference electrode. (**Preston and Shapiro 1998, p. 146; Jun Kimura, p. 39**)

28. (**B**): The low frequency filter has to be increased to 500 Hz; this effectively attenuates potentiates at a distance. This allows studying a single muscle fiber pair. The single fiber muscle action potentials are only recorded if the muscle action potentials are within 200 to 300 μm. Increased jitter is consistent with a NMJ

TABLE 16.1 *Spontaneous Activity in Needle Electromyography (EMG)*

Potential	Firing Pattern	Frequency (Hz)	Source
End-plate activity	Irregular	20–40	Miniature end-plate potential
Positive waves	Regular	0.5–10	Muscle fiber
Fibrillation	Regular	0.5–10	Muscle fiber
Fasciculations	Irregular	0.1–10	Motor unit
Myotonia	Irregular	20–150	Muscle fiber
Myokymia	Regular	1–5	Motor unit
Complex repetitive discharge (CRD)	Regular	5–100	Motor unit
Cramp	Irregular	20–150	Motor unit
Neuromyotonia	Irregular	150–250	Motor unit

disorder. A normal single fiber examination of a clinically weak muscle effectively rules out the diagnosis of myasthenia gravis. Single fiber needle is a modified concentric needle electrode with a smaller leading surface area than a conventional concentric needle electrode. In some patients with ocular myasthenia all study results, including single fiber EMG is normal. (**Preston and Shapiro 1998, p. 508**)

29. (D): Spontaneous activity (defined as activity lasting for more than <3 seconds) is abnormal except for potentials that occur in the muscle end-plate region (i.e., the NMJ). It is found near the center of the muscle belly. Patients often perceive a deep, burning unpleasant sensation when the needle is placed in the end-plate region. Two types of spontaneous activity occur there: end-plate noise and end-plate spike. It is of utmost importance to properly identify these potentials so as not to mistake them for pathologic spontaneous activity (see Table 16.1). (**Preston and Shapiro 1998, p. 182**)

References

1. Preston DC, Shapiro BE. *Electromyography and neuromuscular disorders*, 1st ed. Butterworth-Heineman; 1998.
2. Jun Kimura. *Electrodiagnosis in diseases of nerve and muscle*, 2nd ed. F.A. Davis Company; 1989.

CHAPTER 17

Radiculopathies and Motor Neuron Disease

QUESTIONS

1. **The following statements regarding radiculopathy are correct except:**
   A. Pain and paresthesias radiate in a dermatomal pattern
   B. Paraspinal muscle spasm is absent
   C. Sensory abnormalities are usually vague
   D. Muscle weakness is mild if only one root involved

2. **The most common nerve root that is affected in a cervical radiculopathy is:**
   A. C5
   B. C6
   C. C7
   D. C8

3. **All statements regarding electromyographic (EMG) evaluation of radiculopathy are correct except:**
   A. If the lesion is acute the EMG study may be normal
   B. If the sensory nerve root is predominantly affected the EMG study will be normal
   C. Paraspinal muscles are always abnormal
   D. Fibrillations in the paraspinal muscles do not imply radiculopathy

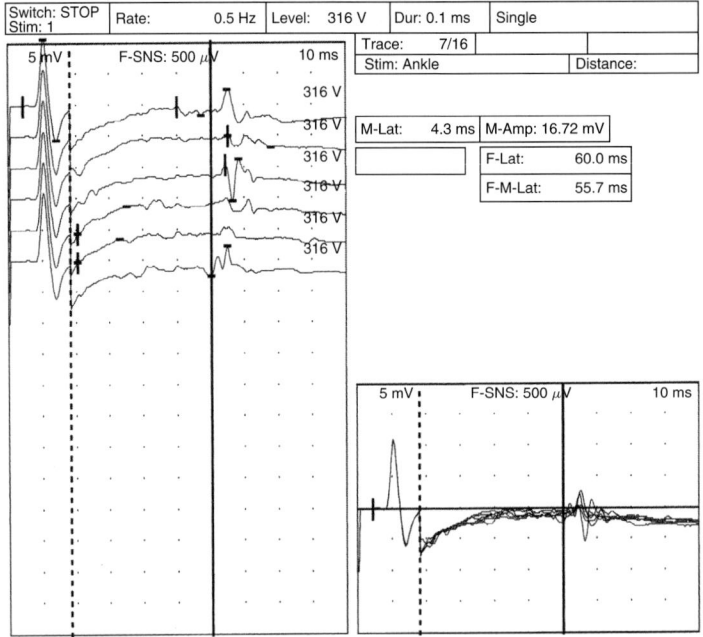

Figure 17.1. Tibial F-wave

4. A 45-year-old female patient presented with lower back pain radiating in the right leg. A nerve conduction study should be conducted on all of the following nerves except:
   A. Sural nerve
   B. Medial plantar nerve
   C. Peroneal nerve
   D. Tibial nerve (see Fig. 17.1)

5. The differential diagnosis now consists of all except:
   A. Lumbosacral radiculopathy
   B. Lumbosacral plexopathy
   C. Peroneal neuropathy
   D. Tibial neuropathy

6. A needle EMG of medial gastrocnemius, tibialis anterior, vastus medialis gluteus maximus, and tensor facia lata is performed. The results are as follows:

   Medial Gastrocnemius Muscle: Fibrillations 2+, positive sharp 1+, polyphasic motor units 15% of units; recruitment decreased
   Tibilis Anterior Muscle: Fibrillations none, positive sharp none, polyphasic motor units none; recruitment normal
   Vastus Medialis Muscle: Fibrillations none, positive sharp none, polyphasic motor units none; recruitment normal
   Tensor Fascia Lata Muscle: Fibrillations none, positive sharp none, polyphasic motor units none; recruitment normal
   Gluteus Maximus Muscle: Fibrillations 1+, positive sharp 1+, polyphasic motor units 15% of units; recruitment decreased
   Lumbosacral Paraspinal Muscles: Fibrillations none, positive sharp none, polyphasic motor units none; recruitment normal

The final diagnosis is:
A. L4 radiculopathy
B. L5 radiculopathy
C. S1 radiculopathy
D. Lumbosacral plexopathy

Refer to the following clinical scenario for Questions 7 through 11:
Clinical Scenario: A 44-year-old right handed man presents to your EMG laboratory for the evaluation of left hand weakness for 3 months.

7. **All of the following tests should be performed except:**
   A. Left median and ulnar sensory nerve conduction study
   B. Left median and ulnar distal motor latency
   C. Left median and ulnar F-wave latency
   D. Sympathetic skin response

8. **The nerve conduction studies come back as:**

   Left median antidromic thumb sensory latency—2.5 ms
   Ulnar antidromic thumb sensory latency—2.0 ms
   Left median distal motor latency—2.8 ms, amplitude 6.8 mV (no conduction block)
   Left ulnar distal motor latency—2.4 ms, amplitude 4.5 mV (no conduction block)
   Left median F-wave latency—29 ms
   Left ulnar F-wave latency—30 ms

   Interpretation of the above findings is:
   A. Left median sensory response is abnormal
   B. Left median motor response is abnormal
   C. Left ulnar motor response is abnormal
   D. Left ulnar sensory response is abnormal

9. **The patient is examined with needle EMG. The muscles sampled and the findings are as follows:**

   Left Deltoid Muscle: Fibrillations none, positive sharp none, fasciculation none, polyphasic motor units 15% of units; recruitment normal
   Left Tricep Muscle: Fibrillations 2+, positive sharp 1+, fasciculation 2+, polyphasic motor units 15% of units; recruitment decreased
   Left First Dorsal Interosseous Muscle: Fibrillations 1+, positive sharp 2+, fasciculation 1+, polyphasic motor units 15% of units; recruitment decreased
   Left Pronator Teres Muscle: Fibrillations none, positive sharp 1+, fasciculation 1+, polyphasic motor units 5% of units; recruitment normal

   Based on these findings the differential diagnosis includes all of the following except:
   A. Left brachial plexopathy
   B. Left cervical radiculopathy
   C. Motor neuron disease
   D. Myasthenia gravis

10. You will proceed with all the following tests except:
    A. Cervical paraspinal muscles testing with needle EMG
    B. Repetitive stimulation test
    C. Needle EMG of the leg
    D. Needle EMG of the right arm

11. You are suspecting motor neuron disease now and would like to confirm it. You will expand your needle EMG to include one of the following muscles:
    A. Thoracic paraspinal muscles
    B. Supraspinatus
    C. Iliopsoas
    D. Lumbosacral paraspinal muscles

12. EMG criteria consistent with amyotrophic lateral sclerosis (ALS) are as follows except:
    A. Motor unit potentials are reduced in number and increased in duration and amplitude
    B. Documenting lower motor neuron abnormalities in the thoracic spine is suggestive of ALS
    C. Fasciculation potentials alone are diagnostic of ALS
    D. Acute denervation is seen in tongue and facial muscles in patients with bulbar presentation

13. In patients with mild radiculopathy, motor nerve conduction studies typically show:
    A. Abnormal F-wave latency
    B. Prolonged distal motor latency
    C. Slow conduction velocity
    D. Normal conduction studies

14. In patients with radiculopathy, sensory nerve conduction studies typically show:
    A. Absent sensory response
    B. Prolonged distal sensory latency
    C. Slow sensory conduction velocity
    D. Normal sensory conduction studies

15. A patient had a back injury 6 weeks ago. His symptoms are suggestive of an S1 radiculopathy. Fibrillation and positive sharp waves are seen in S1 radiculopathy of all of the following muscles except:
    A. Tibialis anterior
    B. Gluteus maximus
    C. Medial gastrocnemius
    D. Abductor hallucis brevis

16. The acute denervation pattern seen in the following muscle will help distinguish a C5 radiculopathy from a C6 radiculopathy:
    A. Deltoid
    B. Biceps brachii

C. Supraspinatus
D. Rhomboid major

ANSWERS

1. (B): In radiculopathy, pain and paresthesia radiate in the distribution of a nerve root. This is usually associated with sensory loss and paraspinal muscle spasm. Associated paraspinal muscle spasm commonly limits the range of motion. Each dermatome overlaps widely with the adjacent dermatome. Consequently, it is very unusual to have a severe or dense sensory disturbance. Similarly there is a wide overlap of myotomes, therefore, muscle weakness is mild if only one nerve root is involved. (**Preston and Shapiro 1998, p. 413**)

2. (C): In cervical radiculopathy, C7 is the most common nerve root affected followed by C6 and C8 (see Table 17.1). The pain and paresthesia radiate in the

TABLE 17.1 *Neurologic Signs and Symptoms with Nerve Root Irritation or Damage from Disc Disease*

Root	Disk	Pain	Sensory Findings	Motor Findings	Reflex Changes
C5	C4-5	Neck, shoulder, and anterior arm	Lateral arm	Deltoid, external rotators of arm, forearm flexors	Biceps, brachioradialis
C6	C5-6	Lateral arm and dorsal forearm	Lateral forearm, lateral arm, and first and second digits	Forearm flexion, arm pronation, finger and wrist extension	Biceps, brachioradialis
C7	C6-7	Dorsal forearm	Third and fourth digits	Arm extension, finger and wrist flexors and extensors	Triceps
C8	C7-T1	Medical forearm and hand, fifth digit	Medial forearm and hand, fifth digit	Intrinsic hand muscles	Finger flexor
L4	L3-4	Low back, buttock, anterolateral thigh, anterior leg	Knee and medial leg	Knee extension	Patellar
L5	L4-5	Low back, buttock, lateral thigh, anterolateral calf	Lateral leg, dorsomedial foot, large toe	Thigh adduction, knee flexion and dorsiflexion of foot and toes	None
S1	L5-S1	Low back, buttock, lateral thigh, calf	Lateral foot, sole of foot, small toe	Hip extension, plantar flexion of foot and toes	Achilles

(From Brazis PW, Masdeu JC, Biller J. *Localization in clinical neurology*, 5th ed. Philadelphia: Lippincott Williams & Wilkins; 2007.)

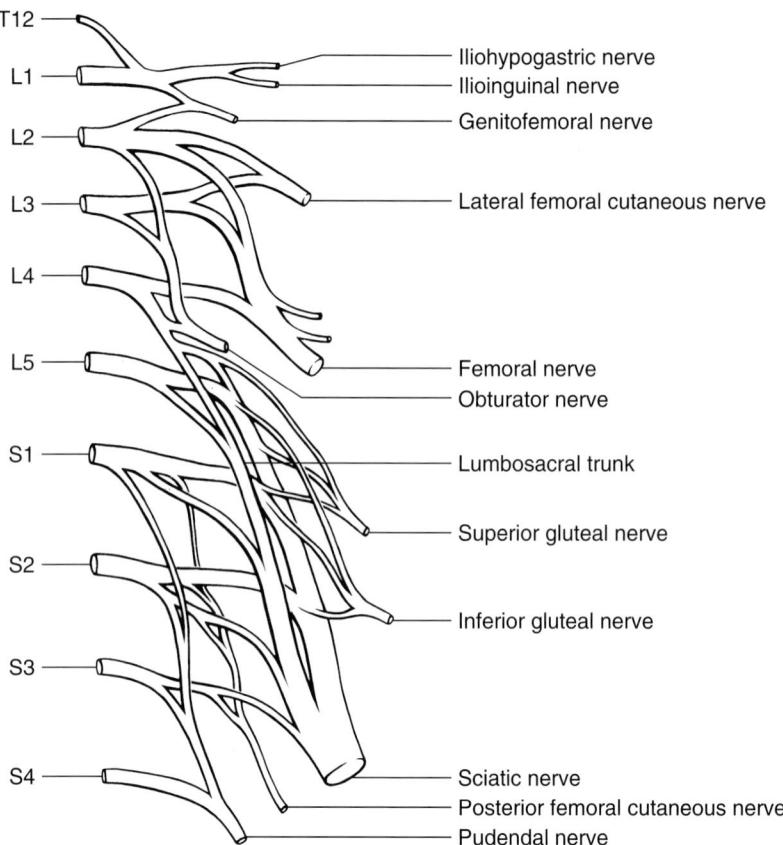

Figure 17.2. The lumbosacral plexus (From Brazis PW, Masdeu JC, Biller J. *Localization in clinical neurology*, 5th ed. Philadelphia: Lippincott Williams & Wilkins; 2007.)

distribution of C7 nerve root (neck shoulder dorsum of forearm and middle finger). As there is a wide overlap of myotomes, even in the case of a severe or complete C7 radiculopathy, the triceps brachii is weak but not paralyzed, retaining some strength from its partial C6 and C8 innervation. Triceps jerk is either decreased or absent in the case of C7 radiculopathy. (**Preston and Shapiro 1998, p. 413**)

3. (C): During the first 10 to 14 days after the onset of an acute radiculopathy, there are no needle EMG abnormalities except for decreased recruitment (see Fig. 17.2). If the sensory nerve root is predominantly affected, the EMG study will be normal. In some cases the paraspinal muscles are completely normal and this happens if there is fascicular sparing of fibers to the dorsal rami or may simply be due to sampling error. Similarly, some patients may not relax during the examination making it difficult to assess the paraspinal muscles. Presence of fibrillations in the paraspinal muscles does not always imply radiculopathy as proximal myopathies, motor neuron disease, botulism, and diabetic polyneuropathy affecting the dorsal rami can cause fibrillations. Similarly patients may have persistent fibrillation in paraspinal muscles after spinal surgery. (**Preston and Shapiro 1998, p. 421**)

4. (B): All the above nerves will help to see if there is radiculopathy while medial plantar nerve will not add much information that is not already obtained by the above nerves.

Her sural sensory latency was normal. Peroneal distal motor latency was 5.5 ms with normal amplitude. Tibial distal motor latency was 5.8 ms with normal amplitude with borderline normal F-wave latency (Fig. 17.1). (**Preston and Shapiro 1998, p. 417**)

5. (C): Muscles need to be evaluated with needle EMG to assess if there is tibial neuropathy versus lumbosacral radiculopathy. Peroneal neuropathy can be excluded from differential when peroneal motor response is normal and there is no conduction block seen across fibular head. (**Preston and Shapiro 1998, p. 311**)

6. (C): The medial gastrocnemius and gluteus maximus are innervated by S1 root and therefore, S1 radiculopathy is the correct answer. Documenting denervation in proximal and distal muscles innervated by the same root is very helpful in confirming the radiculopathy. Some times presence of denervation only in the distal muscles can be confusing as distal polyneuropathy can give the same pattern. It is important to check proximal as distal muscles. Reinnervation, like denervation occurs in proximal before distal muscles. Sometimes proximal muscles are successfully reinnervated and may show denervation only in distal muscle as is seen in this case where paraspinal muscles are normal. (**Preston and Shapiro 1998, p. 420**)

7. (D): In the clinical scenario, the patient presents with left hand weakness. Sympathetic skin response will evaluate the small fiber function and not the motor or sensory large fiber functions. Therefore, all the above studies are appropriate except the sympathetic skin response. (**Preston and Shapiro 1998, p. 424**)

8. (C): Left ulnar motor response shows a normal distal motor and F-wave latency, however, the left ulnar compound motor action potential (CMAP) amplitude is low and is possibly indicative of an axonal problem. All the other values are normal and this effectively rules out carpal tunnel syndrome. (**Preston and Shapiro 1998, p. 238**)

9. (D): Myasthenia gravis does not seem to be a good option. In neuromuscular junction disorders such as myasthenia gravis, mild weakness can be noticed but this degree of acute denervation is not seen. So myasthenia gravis can be taken out of the differential diagnosis. Of note, polyphasic motor units can be seen in normal muscles up to 10% and in deltoid muscle up to 25%. In the findings, the left deltoid and left pronator teres polyphasic motor units are within normal limits. (**Preston and Shapiro 1998, p. 504**)

10. (B): The pattern of EMG abnormality in Question 9 suggests acute and chronic denervation in multiple roots. This can be seen with plexopathy versus radiculopathy versus ALS. The presence of fasciculations is more in favor of a motor neuron disease with involvement of multiple root levels. To ascertain the diagnosis, other limbs and paraspinal muscles will need to be checked.

The needle study of right, left leg, and cervical paraspinal muscles showed acute and chronic denervation pattern in multiple root levels. (**Preston and Shapiro 1998, p. 400**)

11. (A): In the EMG tracings in Question 9, acute and chronic denervation muscles have been previously confirmed. However, cervical and lumbosacral are prone to have wear and turn and can be affected by radiculopathy. Therefore, denervation in thoracic paraspinal muscles in patients with motor neuron disease is important in eliminating the possibility of coexistent cervical and lumbar spinal stenosis mimicking motor neuron disease. Supraspinatus and iliopsoas may show acute and chronic denervation pattern but will not add much to confirm the diagnosis of motor neuron disease. (**Preston and Shapiro 1998, p. 400**)

12. (B): EMG evaluation is very helpful in establishing the diagnosis. Patients with widespread acute and chronic denervation would need testing of relatively fewer muscles while patients with bulbar presentation may need more muscles to be samples. Fasciculation potentials are very common in ALS and help support the diagnosis, however, alone they are not diagnostic unless associated with fibrillation. Acute denervation is seen in tongue and facial muscles in patients with bulbar presentation. Documenting lower motor neuron abnormalities in the thoracic spine is suggestive of ALS. (**Ross 2002, p. 28**)

13. (D): In patients with radiculopathy, nerve conduction studies are typically normal and the electrodiagnosis is established by the needle EMG. The main reason to perform nerve conduction studies is to rule out other conditions, for example, neuropathy or plexopathy as they can mimic some of the same symptoms. Abnormal F-wave responses are seen only in severe radiculopathy when all or most of the motor root fibers are affected. F-wave responses will be normal in a mild radiculopathy, as unaffected motor roots will conduct the F-waves. (**Preston and Shapiro 1998, p. 217**)

14. (D): Sensory studies are the most important part of the nerve conduction studies in the evaluation of radiculopathy. The sensory nerve action potential remains normal in lesions proximal to the dorsal root ganglion. Nearly all radiculopathies damage the root proximal to the dorsal root ganglion. Conversely, lesions distal to the dorsal root ganglion (plexopathy and peripheral nerve) result in decreased sensory nerve action potential (SNAP) amplitude. The presence of a normal SNAP in the same distribution as sensory symptoms should always suggest a lesion proximal to the dorsal root ganglion (radiculopathy). (**Preston and Shapiro 1998, p. 418**)

15. (A): It is not until 3 to 4 weeks that fibrillations develop in the lower leg S1 innervated muscles. All of the above muscles are S1 innervated except tibialis anterior which is an L5>L4 innervated muscle. Sampling gluteus maximus would be helpful to show denervation in a proximal muscle innervated by the S1 root and exclude a sciatic neuropathy. (**Preston and Shapiro 1998, p. 422**)

16. (D): Finding fibrillation in biceps, deltoid, and supraspinatus is consistent with a C5-6 myotomal pattern. At times it can be more challenging to differentiate a C5 from a C6 radiculopathy. If the rhomboids (C5>C4) are abnormal and the pronator teres (C6-7) is normal, then a C5 radiculopathy is more likely than a C6 radiculopathy. (**Preston and Shapiro 1998, p. 421**)

References

1. Preston DC, Shapiro BE. *Electromyography and neuromuscular disorders*, 1st ed. Butterworth-Heineman; 1998.
2. Mark A. Ross. *Continuum*; volume 8, Number 4; August 2002.

CHAPTER 18

Plexopathies and Myopathies

QUESTIONS

1. A patient presents after a motor vehicle accident (MVA) with right upper extremity weakness and sensory loss. On needle electromyogram (EMG), acute denervation pattern is seen in all muscles in the shoulder region and the arm, forearm, and hand except serratus anterior and rhomboids. This pattern is seen with:
   A. Upper trunk plexopathy
   B. Middle trunk plexopathy
   C. Lower trunk plexopathy
   D. Panplexopathy

2. A patient presents after an MVA with right upper extremity weakness and sensory loss. On nerve conduction study, median and radial sensory responses are absent (recorded at thumb) while median and ulnar motor conduction studies and F-responses are normal. Needle EMG shows acute denervation in deltoid and supraspinatus muscles. This pattern is seen with:
   A. Upper trunk plexopathy
   B. Middle trunk plexopathy
   C. Lower trunk plexopathy
   D. Panplexopathy

3. A patient presents after an MVA with right upper extremity weakness and sensory loss. On nerve conduction study, median sensory studies recorded at the middle finger and radial sensory responses at the wrist are absent while median and ulnar motor conduction studies and F-responses are

normal. Needle EMG shows denervation pattern in the pronator teres, triceps, and the flexor carpi radialis. This pattern is seen with:
- A. Upper trunk plexopathy
- B. Middle trunk plexopathy
- C. Lower trunk plexopathy
- D. Panplexopathy

4. A patient presents with a right upper extremity weakness and sensory loss after waking up from coronary bypass graft. On nerve conduction study, ulnar sensory, dorsal ulnar cutaneous and medial antebrachial cutaneous sensory nerve action potential (SNAP), and median and ulnar motor conduction studies and F-responses are absent or abnormal. Radial motor conduction studies are normal. Needle EMG shows denervation pattern in extensor indicis proprius, flexor pollicis longus, and abductor pollicis brevis. This pattern is seen with:
   - A. Upper trunk plexopathy
   - B. Middle trunk plexopathy
   - C. Lower trunk plexopathy
   - D. Panplexopathy

5. A patient presents with progressive right medial hand numbness and weakness with claw-like posture of the hand. On nerve conduction study, ulnar, dorsal ulnar and medial antebrachial cutaneous SNAP are abnormal. The right ulnar and median motor conduction studies and F-responses are absent or abnormal. Radial motor conduction studies are normal. Needle EMG shows a denervation pattern in flexor pollicis longus, flexor carp ulnaris, first dorsal interosseous, and abductor pollicis brevis muscles. This pattern is seen with:
   - A. Upper trunk plexopathy
   - B. Middle trunk plexopathy
   - C. Lateral cord plexopathy
   - D. Thoracic outlet syndrome.

6. A patient presents with the acute onset of severe proximal arm and shoulder pain for the last 2 days. Nerve conduction studies show normal median, ulnar, and radial sensory and motor responses. Needle EMG showed normal first dorsal interossei, abductor pollicis brevis, pronator teres, deltoid, biceps, and triceps muscle findings. The lesion is likely present in the:
   - A. Lower trunk plexopathy
   - B. Axillary nerve
   - C. Suprascapular nerve
   - D. Musculocutaneous nerve

7. All statements regarding needle EMG in myopathy are correct except:
   - A. Examination of proximal and distal muscle should be done
   - B. Muscle biopsy of the muscle, contralateral to the side sampled by the EMG should be done
   - C. Denervating potentials occur frequently in many myopathic disorders
   - D. Contractures are associated with acute denervation pattern on needle EMG

8. **Needle EMG in myopathy is characterized by:**
   A. In myopathy, motor unit territory typically decreases in size as individual muscle fibers drop out
   B. Motor unit action potential (MUAP) amplitude is decreased
   C. MUAP phases can be increased in myopathy
   D. MUAP duration increases in myopathy

9. **Which of the following forms of spontaneous activity in needle EMG is normal?**
   A. Myotonia
   B. Positive sharp waves
   C. Fibrillation potentials
   D. End-plate potentials

10. **A typical EMG finding in polymyositis and dermatomyositis is:**
    A. Motor nerve conduction studies are abnormal
    B. Recruitment pattern is typically decreased
    C. Acute denervation pattern is not seen
    D. Motor unit action potentials are small, short, and polyphasic

11. **The EMG findings seen in inclusion body myositis are characterized by all except:**
    A. Small short MUAPs with polyphasia
    B. Large long MUAPs with polyphasia
    C. Small short and large long MUAPs with polyphasia
    D. Normal study

12. **The EMG finding seen in steroid myopathy that differentiates it from polymyositis is:**
    A. Distal myopathy
    B. Motor studies are normal
    C. Absence of abnormal spontaneous activity
    D. Low amplitude and short-duration MUAPs

13. **All of the following statements are incorrect except:**
    A. In myotonic dystrophy, cooling of the muscle enhances myotonic discharges
    B. Myotonic discharges are generally most prominent in the proximal muscles
    C. In paramyotonia congenita, cooling of the muscle enhances myotonic discharges
    D. In hyperkalemic periodic paralysis, cooling of the muscle enhances myotonic discharges

14. **A 23-year-old woman presents with bilateral shoulder weakness over a period of 7 weeks. She denies any paresthesia or numbness. Nerve conduction study results are as follows:**

    Sural, median and ulnar sensory response—normal
    Left median distal motor latency—2.5, amplitude 19 mV
    Left ulnar distal motor latency—3.0, amplitude 16 mV

Your next test will be:
A. Needle EMG
B. Repetitive nerve stimulation test
C. Nerve biopsy
D. Muscle biopsy

15. The final diagnosis is:
A. Amyotrophic lateral sclerosis (ALS)
B. Inflammatory myopathy
C. Myasthenia gravis
D. Cervical radiculopathy

ANSWERS

1. **(D):** A complete brachial plexopathy results in weakness, sensory loss, and decreased or absent reflexes in the entire arm. If the roots remain intact then serratus anterior and rhomboids are usually the only muscles spared because they are innervated by nerves arising directly from the roots. The assessment of these two muscles in often a key, both clinically and electrically, in differentiating a plexopathy from radiculopathy. Cervical root avulsion will result in a denervation pattern in the above two muscles and portends a poor prognosis. (**Preston and Shapiro 1998, p. 435**)

2. **(A):** In upper trunk lesions, the biceps and brachioradialis tendon reflexes are depressed or absent, but the triceps reflex is spared. In addition to nerve conduction studies, median and radial sensory responses are absent or reduced while median and ulnar motor conduction studies and F-responses are normal. Needle EMG abnormalities may involve the deltoid, biceps, brachioradialis, supraspinatus, and infraspinatus. The rhomboids, serratus anterior, and cervical paraspinal muscles are spared. (**Preston and Shapiro 1998, p. 443**)

3. **(B):** Middle trunk lesions are very rare but may affect the median SNAP, especially when recording the middle finger. Because the middle trunk is formed directly of C7 root, it mimics C7 radiculopathy. The radial SNAP may be absent as well. The triceps jerk is depressed or absent. On nerve conduction study, median sensory studies recorded at the middle finger and radial sensory responses at the wrist are absent or reduced, while median and ulnar motor conduction studies and F-responses are normal. The median nerve sensory result will be abnormal when recorded at the thumb or index finger. Needle EMG abnormalities involve C7 innervated muscles, for example, triceps, pronator teres, and flexor carpi radialis. The cervical paraspinal muscles are spared. (**Preston and Shapiro 1998, p. 444**)

4. **(C):** Lower trunk is formed from C8 to T1 roots. Lower trunk lesions affect the ulnar sensory, dorsal ulnar, and medial antebrachial cutaneous SNAPs (see Fig. 18.1). Because the median and ulnar innervated hand muscles are derived from the lower trunk, their respective motor conduction studies and F-responses may be abnormal. The radial SNAP and motor conduction studies are normal.

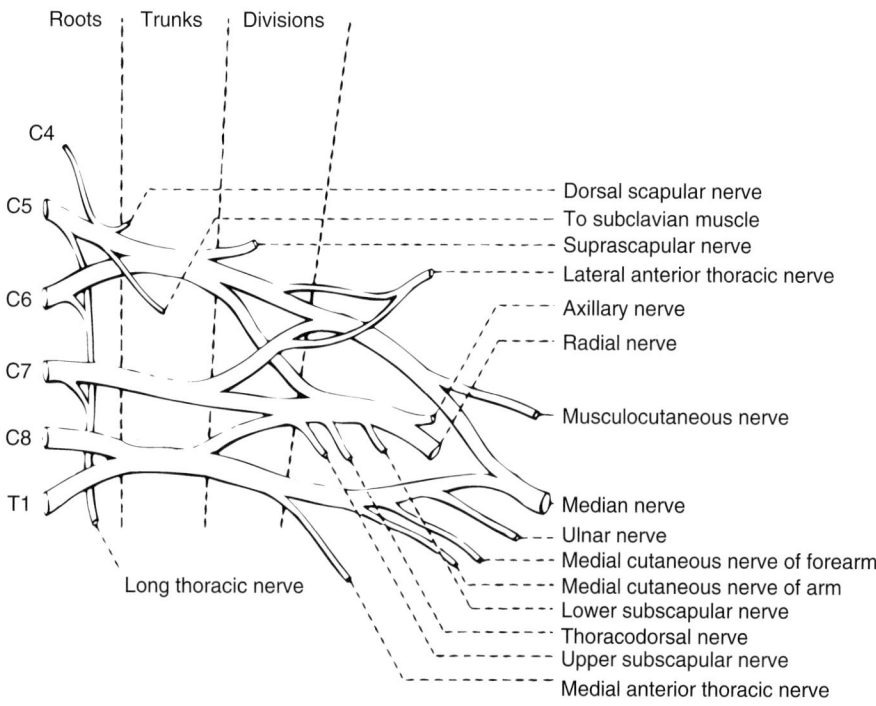

Figure 18.1. The brachial plexus (From Brazis PW, Masdeu JC, Biller J. *Localization in clinical neurology*, 5th ed. Philadelphia: Lippincott Williams & Wilkins; 2007.)

The vast majority of these lesions are axonal rather than demyelinating and motor conduction studies may be normal or may show low amplitude. Needle EMG abnormalities may show abnormalities in all ulnar innervated muscles as well as in the median and radial innervated muscles that contain C8-T1 fibers. In pure lower trunk plexopathies, there are no reflex abnormalities. (**Preston and Shapiro 1998, p. 444**)

5. (**D**): True neurogenic thoracic outlet syndrome (TOS) is actually a lower trunk plexopathy. In this entrapment neuropathy, T1 fibers tend to be preferentially affected, leading to a distinctive pattern on nerve conductions and EMG. The median and ulnar motor distal latencies and conduction velocities may be slightly slowed. The median SNAP is normal because the median sensory fibers do not travel through the lower trunk but rather through upper and middle trunks that are not involved in neurogenic TOS. However, ulnar sensory response is abnormal because the ulnar sensory fibers travel through the lower trunk. Needle EMG abnormalities may show abnormalities in median more than ulnar innervated C8-T1 innervated muscles. In pure lower trunk plexopathies, there are no reflex abnormalities. (**Preston and Shapiro 1998, p. 444**)

6. (**C**): Suprascapular nerve comes off the upper trunk of the brachial plexus (see Fig. 18.2). The nerve runs posteriorly under the trapezius, passing through suprascapular notch of the scapula to enter the supraspinus fossa. Suprascapular entrapment most commonly occurs at the suprascapular notch. The pain is typically deep and boring, occurring along the superior aspect of scapula and radiating to the shoulder. Neuralgic amyotrophy often presents with severe proximal arm and shoulder pain and later weakness. In some cases, the

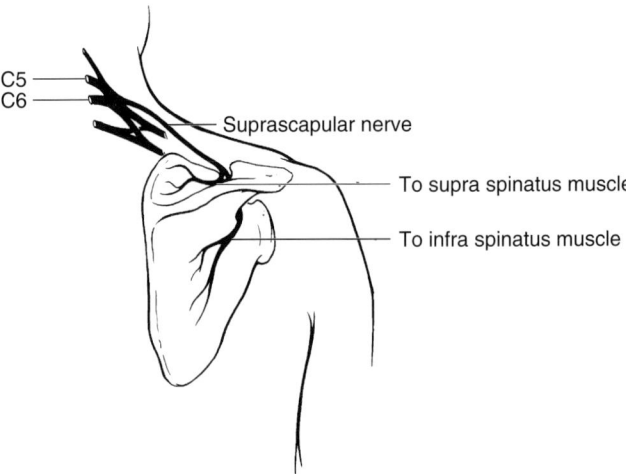

Figure 18.2. The suprascapular nerve (From Brazis PW, Masdeu JC, Biller J. *Localization in clinical neurology*, 5th ed. Philadelphia: Lippincott Williams & Wilkins; 2007.)

suprascapular nerve may be preferentially involved. In lesions at the suprascapular notch, both supraspinatus and infraspinatus are abnormal. In a spinoglenoid lesion only infraspinatus is involved. (**Preston and Shapiro 1998, p. 456**)

7. (**D**): The needle EMG examination must be individualized based on the distribution of patient's symptoms. Overall, examination of proximal and distal muscle should be done. Sampling the paraspinal muscles is often very useful as most myopathies affect the proximal muscles. As the needle EMG may induce transient inflammatory changes in the muscles, muscle biopsy of the muscle, contralateral to the side sampled by the EMG, should be done. Denervating potentials occur frequently in many myopathic disorders. They are likely secondary to muscle fiber necrosis, inflammation, or irritation near the site of neuromuscular junction (NMJ). The presence of denervating potentials often suggests the diagnosis of inflammatory myopathy. Contractures are associated with complete absence of any electroencephalogram (EEG). (**Preston and Shapiro 1998, p. 528**)

8. (**D**): MUAP duration is the most important parameter in myopathy. MUAP duration closely reflects the total number of muscle fibers in a motor unit. As in myopathy, motor unit territory typically decreases in size as individual muscle fibers drop out, the MUAP duration characteristically decreases. MUAP amplitude is usually decreased but can be normal or increased if the needle electrode is placed near reinnervated fibers. MUAP phases are increased in myopathy (>4 phases). The number of phases is primarily a measure of synchrony and polyphasia can be seen with myopathy and neuropathy. (**Preston and Shapiro 1998, p. 529**)

9. (**D**): Spontaneous activity (defined as activity lasting for >3 seconds) is abnormal except for potentials that occur in the muscle end-plate region (i.e., the NMJ). It is found near the center of the muscle belly. Patients often perceive a deep, burning unpleasant sensation when the needle is placed in the end-plate

region. Two types of spontaneous activity occur there, end-plate noise and end-plate spike. It is of utmost importance to properly identify these potentials so as not to mistake them for pathologic spontaneous activity. **(Preston and Shapiro 1998, p. 182)**

10. **(D):** In polymyositis and dermatomyositis, motor and sensory nerve conduction studies are normal. EMG typically shows prominent spontaneous activity (fibrillations, positive sharp waves, and complex repetitive discharges). In acute and subacute cases, MUAPs are small, short, and polyphasic with early recruitment. Fibrillations are typically present in 45% to 74% of patients. The diagnostic yield increased as more muscles were sampled. Most commonly, these fibrillations are seen in the paraspinal muscles (94%) followed by the proximal shoulder and hip muscles (64% to 76%). **(Preston and Shapiro 1998, p. 531)**

11. **(D):** The electrophysiology often complicates the diagnosis of inclusion body myositis (IBM). A subset of patients demonstrates a mild sensory or sensorimotor polyneuropathy on nerve conduction. In addition, needle EMG is often confusing, prominent denervating potentials are common. The associated MUAP findings fall into one of three separate groups:

a. Small short MUAPs with polyphasia
b. Large, long MUAPs with polyphasia
c. Small, short and large long MUAPs with polyphasia

Because of the heterogeneous profile of IBM, many authorities have commented that the combination of neuropathic and myopathic findings on EMG should suggest IBM. **(Preston and Shapiro 1998, p. 532)**

12. **(C):** Among drug-induced myopathies, steroid myopathy is the most common. The risk of steroid myopathy increases with dose and duration of use. It is typically a proximal myopathy, preferentially affecting the hip girdle muscles. Motor and sensory nerve conduction studies are normal. The needle EMG is often normal or show low amplitude, short-duration MUAPs in proximal muscles. Typically absence of abnormal spontaneous activity in steroid myopathy, helps to differentiate it from polymyositis. **(Preston and Shapiro 1998, p. 532)**

13. **(C):** In some myotonic disorders (paramyotonia congenital and myotonia congenital), cooling of the muscle can be used to enhance myotonic discharges. Cooling of the muscle is done in ice water to 20°C. Transient dense fibrillation potentials appear with cooling. All myotonic discharges disappear below 20°C. This cooling effect is not seen in myotonic dystrophy or hyperkalemia periodic paralysis. Myotonic discharges are generally most prominent in the distal muscles. A recently recognized dystrophic myotonic muscle disorder has been found to have more myotonic discharges in the proximal muscles than in distal muscles. **(Preston and Shapiro 1998, p. 550)**

14. **(A):** The most appropriate next test is needle EMG. This will help assess if this is myopathy versus neuropathic process (i.e., plexopathy or radiculopathy). Nerve or muscle biopsy may be needed but jumping to these invasive tests is not appropriate.

The needle EMG results are as follows:

Left Deltoid Muscle: Fibrillations 2+, positive sharp 1+, small polyphasic motor units 35% of units; recruitment early

Left Triceps Muscle: Fibrillations 1+, positive sharp 1+, small polyphasic motor units 15% of units; recruitment early

Left Biceps Muscle: Fibrillations 1+, positive sharp 3+, small polyphasic motor units 15% of units; recruitment early

Left Pronator Teres Muscle: Fibrillations none, positive sharp none, polyphasic motor units none; recruitment normal

Left First Dorsal Interosseous Muscle: fibrillations none, positive sharp none, polyphasic motor units none; recruitment normal

Left Abductor Policis Brevis Muscles: fibrillations none, positive sharp none, polyphasic motor units none; recruitment normal

15. (B): The EMG pattern suggests myopathy with small polyphasic motor units and early recruitment while the presence of acute denervation pattern is consistent with inflammatory myopathy. In general, motor units in myopathy are small, short in duration, and polyphasic in comparison to neuropathic processes where the motor units are large, long in duration and polyphasic. The early recruitment pattern results from loss of muscle fibers and the remaining muscle fibers try to compensate for the loss of other fibers and result in increased-interference pattern.

References

1. Preston DC, Shapiro BE. *Electromyography and neuromuscular disorders*, 1st ed. Butterworth-Heineman; 1998.

CHAPTER 19

Electromyographic Findings in Common Neuromuscular Disorders

QUESTIONS

1. **A patient presents with a right wrist drop. The acute denervation pattern can be seen in all except:**
   A. Abductor policis brevis
   B. Brachioradialis
   C. Deltoid
   D. Extensor indicis proprius

2. **All statements regarding radial neuropathy are correct except:**
   A. The superficial radial sensory nerve is easy to stimulate and record
   B. Most cases of posterior interosseous neuropathy are purely demyelinating in nature and a conduction block is demonstrated
   C. Radial neuropathy at the spiral groove shows a conduction block with stimulation of the radial nerve proximal to the spiral groove
   D. Posterior interosseous neuropathy results in a normal superficial radial sensory nerve action potential (SNAP)

3. **Acute denervation pattern is seen in all muscles in radial neuropathy at the spiral groove except:**
   A. Brachioradialis
   B. Supinator

C. Anconeus
D. Extensor carpi radialis

4. A patient presents with left foot drop from a common peroneal neuropathy at the knee. All the muscles show acute denervation except:
   A. Tibialis anterior
   B. Peroneus longus
   C. Extensor hallucis longus
   D. Tibialis posterior

5. A patient presents with left foot drop with peroneal neuropathy at the fibular head. All the muscles will show acute denervation except:
   A. Short head of biceps femoris
   B. Peroneus longus
   C. Extensor hallucis longus
   D. Tibialis anterior

6. A patient presents with left foot drop from a common peroneal neuropathy in the thigh. All the muscles will show acute denervation except:
   A. Long head of biceps femoris
   B. Short head of biceps femoris
   C. Extensor hallucis longus
   D. Tibialis anterior
   E. Peroneal longus

7. A patient with a femoral neuropathy below the inguinal ligament shows a denervation pattern in all muscles except:
   A. Rectus femoris
   B. Iliopsoas
   C. Vastus medialis
   D. Vastus lateralis

8. Intrinsic foot muscle atrophy may occur in all of the following except:
   A. Proximal tibial neuropathy
   B. Tarsal tunnel syndrome (TTS)
   C. Peroneal neuropathy
   D. L5-S1 Radiculopathy

9. All statements about the needle electromyogram (EMG) in botulism are correct except:
   A. The needle EMG of botulism is normal
   B. Botulinum toxin is a potent neuromuscular junction (NMJ) blocker resulting in effective chemo-denervation
   C. Motor unit action potentials (MUAPs) may be short and polyphasic
   D. Single fiber EMG shows increased jitter and blocking

10. Congenital myasthenic syndrome is associated with all except:
    A. These disorders are not immune mediated
    B. Usually present shortly after birth or early childhood

C. Single fiber EMG is normal
D. Some patients display a decremental response on slow repetitive nerve stimulation (RNS)

ANSWERS

1. **(A):** See Figure 19.1. The anatomic localization of wrist drop can be posterior interosseous neuropathy, radial neuropathy at the spiral groove, radial neuropathy in the axilla, posterior cord brachial plexopathy, and C7 radiculopathy. These lesions will not result in an acute denervation pattern in the abductor policis

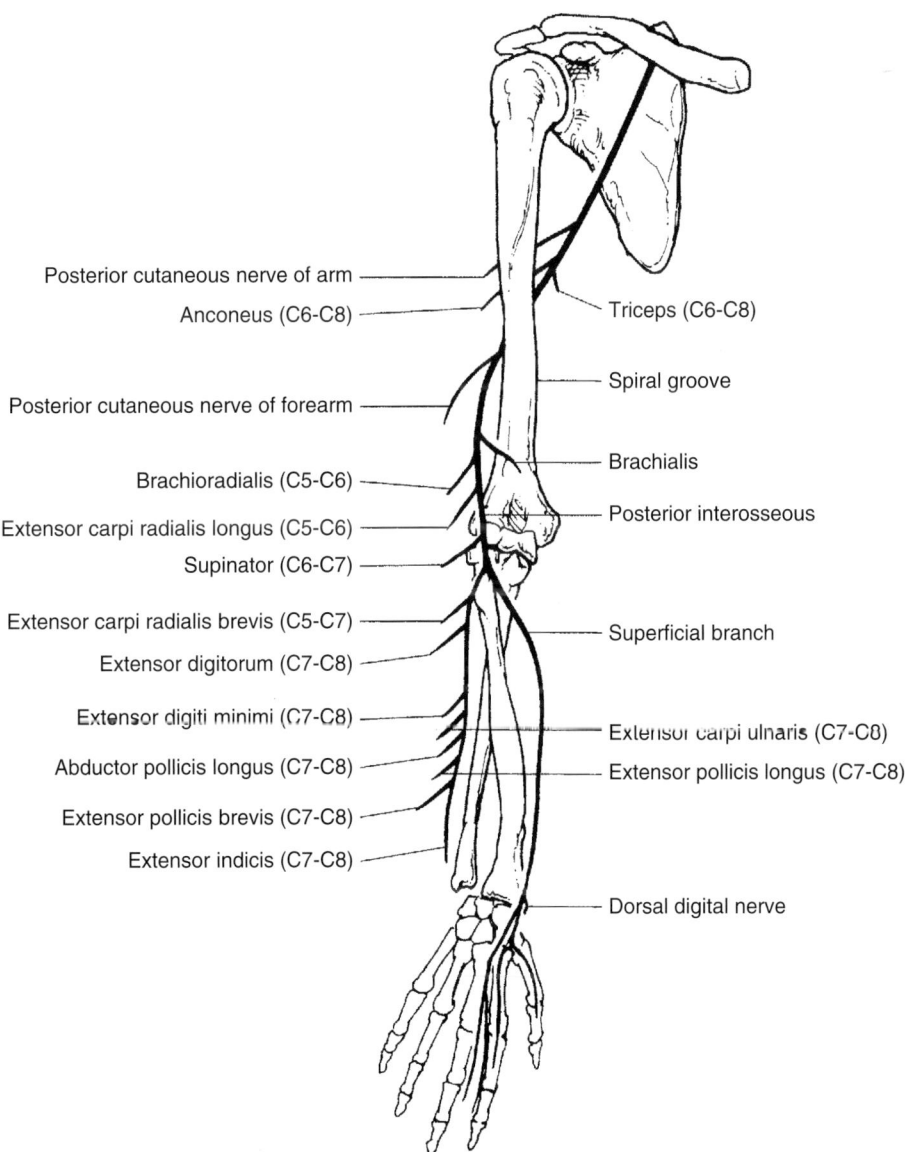

Figure 19.1. The radial nerve (From Brazis PW, Masdeu JC, Biller J. *Localization in clinical neurology*, 5th ed. Philadelphia: Lippincott Williams & Wilkins; 2007.)

brevis. However, acute denervations can be seen in the other three options with wrist drop depending where the lesion is. (**Preston and Shapiro 1998, p. 294**)

2. (**B**): In cases of radial neuropathy at the spiral groove, conduction block is seen with stimulation of the radial nerve proximal to the spiral groove. The relative drop in proximal to distal compound muscle action potential (CMAP) amplitude gives some indication of the proportion of fibers blocked. Most cases of posterior interosseous neuropathy are purely axonal in nature and a conduction block is usually not seen. A normal superficial radial sensory response is seen in posterior interosseous neuropathy because the superficial radial sensory nerve comes off before the posterior interosseous nerve. (**Preston and Shapiro 1998, p. 296**)

3. (**C**): In radial neuropathy at the spiral groove acute denervation is seen in the brachioradialis, long head of the extensor carpi radialis, and supinator muscles. The branch to the anconeus comes off proximal to the spiral groove. If the lesion is at the axilla, the triceps and anconeus will be involved and acute denervation will be seen in anconeus. (**Preston and Shapiro 1998, p. 299**)

4. (**D**): Common peroneal neuropathy above the fibular head results in weakness and acute denervation pattern in the tibialis anterior, extensor hallucis longus, and peroneal longus muscles. However, it will not affect the tibialis posterior as it is a tibial innervated muscle. The only branch of the common peroneal nerve above the knee is to the short head of the biceps femoris. The presence of denervation pattern in the biceps femoris muscle indicates involvement of the common peroneal nerve above the knee. (**Preston and Shapiro 1998, p. 314**)

5. (**A**): See Figure 19.2. Peroneal neuropathy at the fibular head results in a weakness and acute denervation pattern in the tibialis anterior, extensor hallucis longus, and peroneal longus muscles. The only branch of common peroneal nerve above the knee is to biceps femoris. The presence of denervation pattern in biceps femoris indicates involvement of the common peroneal nerve above the knee and above the branch to the short head of the biceps femoris muscle, which will not be seen in this case because the lesion is at the fibular head. (**Preston and Shapiro 1998, p. 314**)

6. (**A**): Common peroneal neuropathy in the knee results in a weakness and acute denervation pattern in the tibialis anterior, extensor hallucis longus, and peroneal longus. The only branch of the common peroneal nerve above the knee is to the short head of the biceps femoris. The presence of denervation pattern in the short head of the biceps femoris indicates involvement of the common peroneal nerve above the knee. However the long head of biceps femoris is a tibial nerve innervated muscle and denervation will not be seen because the lesion is in common peroneal nerve. (**Preston and Shapiro 1998, p. 314**)

7. (**B**): Isolated lesions of femoral nerve are not often seen in EMG findings. Quadriceps muscles are supplied by the femoral nerve and will show indications of denervation in a femoral neuropathy. The branch to iliopsoas comes off from the femoral nerve above the inguinal ligament. A lesion of femoral nerve below the inguinal ligament will not affect the iliopsoas muscle. (**Preston and Shapiro 1998, p. 324**)

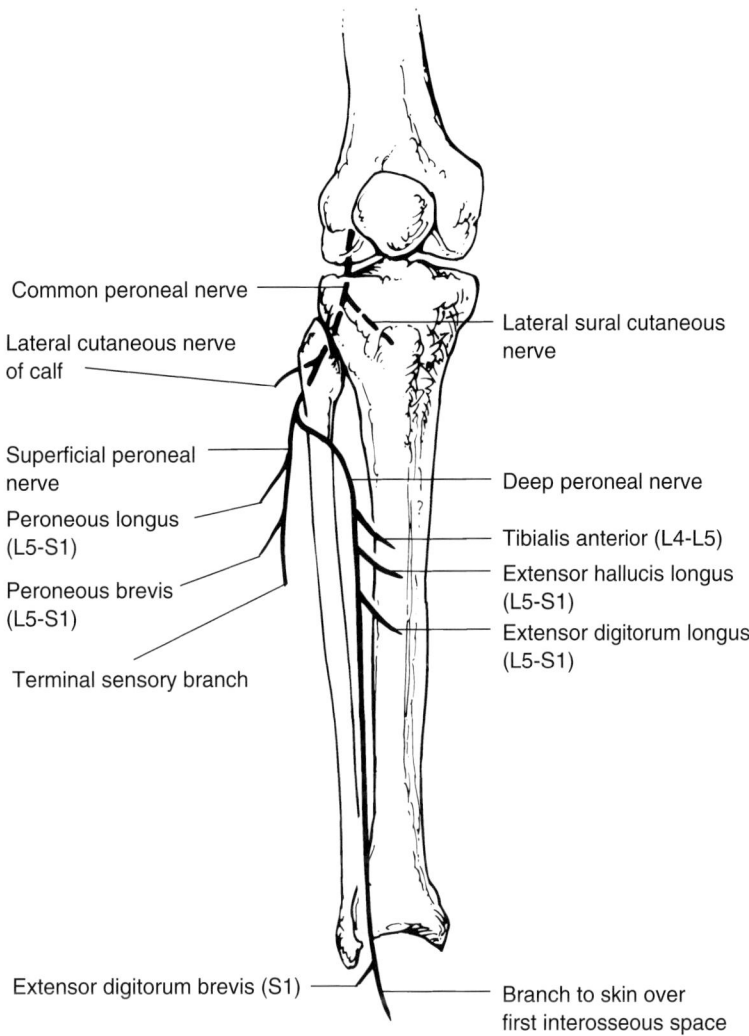

Figure 19.2. The peroneal nerve (From Brazis PW, Masdeu JC, Biller J. *Localization in clinical neurology,* 5th ed. Philadelphia: Lippincott Williams & Wilkins; 2007.)

8. **(C):** Intrinsic foot muscle atrophy can occur in TTS, a proximal tibial neuropathy, and L5-S1 radiculopathy. However a peroneal mononeuropathy will not affect the intrinsic foot muscles. TTS is caused by hypertrophy of the flexor retinaculum from repetitive use (akin to carpal tunnel syndrome [CTS]). A Tinel sign over the tarsal tunnel may help in identifying a TTS, but the sign is nonspecific. (**Preston and Shapiro 1998, p. 329**)

9. **(A):** The needle EMG of botulism is interesting. Fibrillation and positive wave are seen. Botulinum toxin is a potent NMJ blocker resulting in effective chemo-denervation. Similar to other NMJ disorders MUAPs may be normal or short and polyphasic, similar to myopathic MUAPs. Depending on the severity, recruitment may be normal, early, or reduced. Reduced recruitment is seen when every muscle fiber of a motor unit is blocked by the botulinum toxin, effectively reducing the number of motor units. Likewise single fiber EMG shows increased jitter and blocking signifying the underlying NMJ dysfunction. (**Preston and Shapiro 1998, p. 512**)

10. (C): Congenital myasthenic syndromes are a group of exceptionally rare disorders caused by inherited defect in NMJ transmission. These disorders are not immune mediated and are not associated with autoantibodies in the blood. These disorders usually present shortly after birth or in early childhood. Single fiber EMG findings are usually abnormal. Some patients display a decremental response on slow RNS (3-Hz). (**Preston and Shapiro 1998, p. 512**)

References

1. Preston DC, Shapiro BE. *Electromyography and neuromuscular disorders*, 1st ed. Butterworth-Heineman; 1998.

# SECTION 6

Sleep Medicine

CHAPTER 20

General Principles of Sleep Medicine

QUESTIONS

1. **In which structure is rapid eye movement (REM) sleep generated?**
   A. Cortex
   B. Medulla
   C. Pons
   D. Thalamus

2. **Stage 2 sleep is defined by:**
   A. First K-complex or sleep spindle in the first half of the epoch
   B. Vertex waves
   C. 20% to 50% Delta activity
   D. >50% Delta activity

3. **The K-complex is characterized by all except:**
   A. A K-complex must be at least 0.5 seconds
   B. The K-complex is first apparent at approximately age 5 months
   C. K-complex is an evoked response
   D. A K-complex must be followed by a sleep spindle

4. **The REM sleep is characterized by all except:**
   A. K-complexes
   B. Bursts of eye movements
   C. Lack of eye movements and muscle twitches
   D. Muscle atonia

5. **Obstructive sleep apnea is characterized by all except:**
   A. Is worse in REM stage
   B. Is relatively less prominent in non–rapid eye movement (NREM) stage
   C. Apnea and hypopnea index of >8
   D. Apnea and hypopnea index of >5

6. **REM sleep behavior disorder is associated with:**
   A. Atonia during sleep
   B. Acting out their dreams
   C. Apnea/hypopnea index of >5
   D. Is common in females than males by a ratio of 2:1

7. **Symptoms of narcolepsy include all of the following except:**
   A. Sleep paralysis
   B. Cataplexy
   C. Hypnogoggic and hypnopompic hallucinations
   D. Acting out the dreams

8. **Maintenance of wakefulness test is useful for all of the following:**
   A. Measuring the ability of the patient to remain awake
   B. Diagnosing narcolepsy
   C. Assessing the response to stimulants
   D. Assessing patient's fitness to drive or to fly an airplane

9. **All statements concerning narcolepsy are correct except:**
   A. REM sleep occurs within minutes of falling asleep
   B. Sleep paralysis and cataplexy appears to be related to muscle atonia of REM sleep
   C. Hypnogoggic hallucination often occurs just as the narcoleptic passes from REM to and out of REM stage
   D. Frequent nighttime awakenings are not seen with narcolepsy

10. **A multiple sleep latency test (MSLT) diagnostic of narcolepsy is characterized by all of the following except:**
    A. Sleep onset REM in <18 minutes in two of the four trials
    B. Sleep onset latency of <8 minutes
    C. Sleep onset REM within 15 minutes in two of the four trials
    D. Some patients with unequivocal narcolepsy and cataplexy do not have two or more sleep onset REM even with repeated studies

11. **Alpha intrusions are seen in:**
    A. REM sleep if alpha activity is superimposed on normal activity of REM sleep
    B. Narcolepsy
    C. REM behavior disorder
    D. Connection with chronic pain syndrome

12. **Apnea and hypopnea are characterized by all except:**
    A. Apnea is defined as a complete sensation of airflow
    B. Hypopnea is a partial decrease in airflow
    C. Apnea/hypopnea must last 10 seconds or longer with a clear decrease >50% from baseline
    D. Apnea/hypopnea associated with oxyhemoglobin desaturation of >2%

13. **Stage 1 sleep is defined by:**
    A. First K-complex or sleep spindle in the first half of the epoch
    B. Vertex waves
    C. 20% to 50% Delta activity
    D. >50% Delta activity

14. **Stage slow wave sleep is defined by:**
    A. First K-complex or sleep spindle in the first half of the epoch
    B. Vertex waves
    C. >20% Delta activity
    D. Lack of muscle activity

15. **The diagnostic criteria for periodic limb movements of sleep (PLMS) is:**
    A. Usually movements of the arms with rare leg movements
    B. PLMS are scored only if at least two occur in sequence
    C. PLMS are scored only if at least three occur in sequence
    D. PLMS are scored only if at least four occur in sequence

16. **The biologic clock in humans is in which nucleus?**
    A. Supraoptic nucleus
    B. Suprachiasmatic nucleus (SCN)
    C. Superior colliculus nucleus
    D. Raphe magnus nucleus

17. **All statements regarding age-related sleep changes in elderly are correct except:**
    A. The amount of stage 1 sleep and of REM sleep decrease
    B. Daytime drowsiness and nap frequency tend to increase in older adults
    C. Many elderly no longer have enough high-amplitude delta waves to allow stage 3 or 4 sleep to be scored
    D. It is more difficult for elderly to have continuous restful sleep at night

18. **All statements regarding age-related sleep changes in children are correct except:**
    A. The sleep spindles become synchronous by age 3
    B. The amount of REM sleep declines to 25% of total sleep time at age 5
    C. NREM to REM cycle time increases to 80 to 90 minutes by age 5
    D. Slow wave sleep increases to approximately 50% of total NREM sleep by approximately 1 year

19. This nucleus may play a key role in controlling the state change between NREM and REM sleep:
    A. Supraoptic nucleus
    B. SCN
    C. Superior colliculus nucleus
    D. Raphe magnus nucleus

20. The diagnostic criteria for restless leg syndrome (RLS) includes all of the following except:
    A. Paresthesias or dysesthesias of the limbs associated with a desire to move the limbs
    B. Motor restlessness
    C. Worsening of symptoms in the evening and at night
    D. Kicking or twitching of the limbs observed by the bed partner

21. Hypocretin (orexin) deficiency is associated with which of the following condition?
    A. Obstructive sleep apnea
    B. Periodic leg movements of sleep
    C. Narcolepsy
    D. RLS

22. Leptin is associated with which of the following conditions?
    A. Obstructive sleep apnea
    B. Periodic leg movements of sleep
    C. Narcolepsy
    D. RLS

23. All statements concerning sleep hygiene are correct except:
    A. Maintain regular times of going to bed and arising from bed
    B. Avoid consuming caffeinated drinks late in the day
    C. Keep the bedroom dark, quiet, and cool
    D. Use a small quantity of alcohol at bedtime to fall asleep

24. All statements regarding insomnia are correct except:
    A. Many insomniacs report having difficulty falling asleep
    B. Patients report of early morning awakenings
    C. Most insomniacs do not feel rested in the morning
    D. Patients report long periods of "resting" in bed without sleeping

25. Ondine's curse is associated with which of the following conditions?
    A. Narcolepsy
    B. Cataplexy
    C. Central alveolar hypoventilation
    D. Obstructive sleep apnea

ANSWERS

1. (C): Transaction experiments by Jouvet demonstrated that REM sleep is generated in the pons. The cholinergic neurons are essential for the generation of REM sleep. Cholinergic cells of the median and dorsolateral pons increase their firing rates during the REM sleep. (**Aldrich 1999, p. 32**)

2. (A): Stage 2 sleep is characterized on electroencephalogram (EEG) by sleep spindles and K-complexes on background of mixed frequencies predominantly in the theta range in the first half of an epoch. Slow eye movements are much less evident during stage 2 sleep than during stage 1 sleep, although they may reappear for short intervals. By convention, a sleep epoch is 30 seconds. (**Aldrich 1999, p. 11**)

3. (D): The K-complex has a voltage maximal over the vertex or less commonly over the vertex or midline frontal scalp. The K-complex must be at least 0.5 seconds in duration and it need not be followed by sleep spindles. There is no voltage criterion for K-complex. The K-complex is first apparent at approximately age 5 months. K-complex is an evoked response, that is, during stage 2 sleep the K-complex can be easily elicited by auditory stimuli, such as the tap of a pencil and other forms of stimulation. (**Aldrich 1999, p. 12**)

4. (A): The REM sleep is characterized by lack of K-complexes and sleep spindles. The REM is divided into phasic and tonic REM. The phasic REM is characterized by periods of muscle twitching with bursts of eye movements. The tonic REM is characterized by lack of eye movements and muscle twitches. The muscle atonia of REM sleep involves the facial, postural muscles of the neck and accessory muscles of respiration. (**Aldrich 1999, p. 14**)

5. (C): Obstructive sleep apnea is a common disorder and is associated with snoring and excessive daytime sleepiness. Obstructive sleep apnea is characterized by apnea and hypopnea index of >5. It is worse on lying on the back and in REM sleep than in NREM sleep stage. (**Aldrich 1999, p. 207**)

6. (B): REM sleep behavior disorder is associated with dread-enacting behaviors and loss or impairment of normal muscle atonia during REM sleep. REM behavior disorder is a major cause of behavioral disturbance during sleep. For unknown reasons, it is more common in men than in women by a ratio of 2:1 to 5:1. The timing and duration of episodes parallel the distribution of REM sleep across the night. (**Aldrich 1999, p. 271**)

7. (D): Narcolepsy is a disorder characterized by brief episodes of irresistible sleep and fall associated with emotional stimuli. Cataplexy is pathognomonic of narcolepsy, with brief episode of bilateral weakness without alteration of awareness, usually brought on by excitement or emotion. Sleep paralysis is a feeling of paralysis that prevents limb movement and eye opening may be accompanied by a sensation of struggling to move. Hypnogoggic and hypnopompic hallucinations are seen in transition to sleep or waking. Acting out the dreams is seen with REM behavior disorder and is not related to narcolepsy. (**Aldrich 1999, p. 154**)

8. (**B**): Maintenance of wakefulness test is a less frequently used variant of MSLT. This is not a test for sleepiness but measures the ability of the patient to remain awake. It should not be used to diagnose narcolepsy. However, it is very helpful in assessing the response to stimulants. It is also helpful to assess patient's fitness to drive or to fly an airplane. (**Daube 2002, p. 496**)

9. (**D**): Narcolepsy is characterized by the occurrence of REM sleep within minutes of falling asleep, called *sleep onset REM*. Sleep paralysis and cataplexy appears to be related to muscle atonia of REM sleep in waking and related to emotional excitement in case cataplexy. Hypnogoggic hallucination often occurs just as the narcoleptic passes from REM to and out of REM stage. In narcolepsy, frequent nighttime awakenings are seen, indicating that the ability to sustain is also impaired. (**Aldrich 1999, p. 159**)

10. (**A**): An MSLT diagnostic of narcolepsy is characterized by classical sleep onset REM in <15 minutes in two of the four trials. The sleep onset latency is decreased and is usually <8 minutes. However, some patients with unequivocal narcolepsy and cataplexy do not have two or more sleep onset REM even with repeated studies. (**Aldrich 1999, p. 167**)

11. (**D**): Alpha intrusions are scored if alpha activity is superimposed on normal activity of NREM sleep. This phenomenon is called *alpha-delta sleep* or *nonrestorative sleep* and is often associated with chronic pain syndrome, for example, fibromyalgia or rheumatoid arthritis. It can also occur in patients with no medical problems. (**Daube 2002, p. 500**)

12. (**D**): Apnea is characterized by a complete sensation of airflow and hypopnea is characterized by a partial decrease in airflow. Apnea/hypopnea must last 10 seconds or longer with a clear decrease >50% from baseline or is associated with oxyhemoglobin desaturation of >3% or arousal. (**Daube 2002, p. 504**)

13. (**B**): Stage 1 sleep is characterized on EEG by low-amplitude mixed frequency EEG. Vertex waves and rudimentary sleep spindles <0.5 seconds in duration may occur toward the end of stage 1. Stage 2 sleep is characterized by sleep spindles and K-complexes on background of mixed frequencies predominantly in the theta range. Slow eye movements are much less evident during stage 2 sleep than during stage 1 sleep, although they may reappear for short intervals. (**Daube 2002, p. 497**)

14. (**B**): Stage slow wave sleep is characterized on EEG by large-amplitude slow waves >20%. Stage 3 being 20% to 50% delta wave and stage 4 being >50% delta activity. With increasing age, overall amplitude of the EEG diminishes. In elderly, considerable activity may still occur in the 0.5 to 2 Hz range, but amplitudes may only reach 50 μV. (**Daube 2002, p. 498**)

15. (D): PLMS are seen in many normal especially older sleepers with usually movements of the leg but rarely of the arms but may be abnormal if PLMS occur at least four times, occur in sequence with the duration of movements being 0.5 to 5 seconds each. PLMS can accompany narcolepsy or idiopathic hypersomnia. More than 40 PLMS per hour have both a sensitivity and specificity of 81% for the diagnosis of RLS. (**Daube 2002, p. 510**)

16. (B): The SCN located in the anterior hypothalamus just above the optic chiasm is the primary locus of circadian pacemaker. Raphe magnus nuclei have ascending projections and influence the control of sleep–wake state. They fire most rapidly in waking state, decrease firing in NREM sleep, and become silent during REM sleep. (**Aldrich 1999, p. 59**)

17. (A): After age 65, the decline in sleep quality that begins during the middle age accelerates. The amount of stage 1 sleep increases while the amount of REM sleep remains the same. Daytime drowsiness and nap frequency tend to increase in older adults. Many elderly no longer have enough high-amplitude delta waves to allow stage 3 or 4 sleep to be scored. It is more difficult for elderly to have continuous restful sleep at night. (**Aldrich 1999, p. 76**)

18. (A): The sleep spindles tend to become synchronized by 18 months and by age 2 all sleep spindles are synchronized. The amount of REM sleep declines to 25% of total sleep time (the adult value) at age 5. NREM to REM cycle time increases to 80 to 90 minutes (the adult value) at age 5. Slow wave sleep increases to approximately 50% of total NREM sleep by approximately 1 year. Then amount of delta sleep begins to decline after the first few years of life and by age 9 delta sleep makes up approximately 22% to 28% of sleep. (**Aldrich 1999, p. 76**)

19. (D): Raphe magnus nuclei have ascending projections and influence the control of sleep–wake state. They fire most rapidly in waking state, decrease firing in NREM sleep and become silent during REM sleep. They resume their firing at the end of REM, before the increase in muscle tone and cessation of pontine-geniculate-occipital spikes, signals the end of REM sleep. (**Aldrich 1999, p. 30**)

20. (D): In most cases the diagnosis of RLS is made clinically, based on four major clinical aspects; paresthesias or dysesthesias of the limbs associated with a desire to move the limbs, motor restlessness, worsening of symptoms in the evening and at night, and worsening of symptoms at rest. Kicking or twitching of the limbs observed by the bed partner is seen with PLMS and not with RLS. (**Aldrich 1999, p. 179**)

21. (C): Through cerebrospinal fluid (CSF) hypocretin-1 measures, it was found that a large majority of "idiopathic" human narcolepsy cataplexy cases are associated with hypocretin ligand deficiency. Postmortem studies in a small number of narcolepsy–cataplexy subjects confirmed the absence of hypocretin production in the brain parenchyma.

Hypocretin (orexin) neurons are specifically localized in the lateral hypothalamus (LH), but have widespread excitatory projections to monoaminergic, cholinergic, and GABAergic neurons in brainstem, basal forebrain, and cortex. Hypocretin neurons are thought to be implicated in maintaining wakefulness and regulating motor functions (locomotion, muscle tone), feeding, energy expenditure, and sympathetic activity. (**Nishino and Kanbayashi 2005, pp. 269–310**)

22. (A): Leptin, a protein hormone is produced by white adipose and placental tissue. It inhibits neuropeptide Y synthesis in the hypothalamus and downregulates food intake. Leptin is also associated with increased energy expenditure. The high prevalence of obstructive sleep apnea in obese people and the established role of leptin as a respiratory stimulant and appetite suppressant suggest that sleep apnea is a leptin-deficient state. Studies have revealed increased serum leptin levels in patients with obstructive sleep apnea. There is a positive correlation between serum leptin and apnea-hypopnea index (AHI). (**Ciftci 2005, pp. 395–401**)

23. (D): Education about the nature of sleep and the reasons for maintaining good sleep habits is usually the best treatment for insomnia, for example, maintain regular times of going to bed and arising from bed, avoid consuming caffeinated drinks late in the day, keep the bedroom dark, quiet and cool, use the bed only for sleep and sexual activity. It is recommended to avoid evening alcohol use. (**Aldrich 1999, p. 145**)

24. (D): Many insomniacs report having difficulty falling asleep, staying asleep and wake up multiple times. They often report of early morning awakenings. Most insomniacs do not feel rested in the morning and feel tired. Patients with sleep-state misperception (not insomnia) often report long periods of "resting" in bed without sleeping. In these patients, there is a marked discrepancy between complaint of insomnia and observation of normal sleep in the laboratory. (**Aldrich 1999, pp. 124, 142**)

25. (C): Central alveolar hypoventilation often called *Ondine's curse* is a rare disorder of automatic ventilation that may be congenital or acquired. It causes prolonged apneas and periods of hypoventilation that are worse during sleep. In severe cases, cyanosis and apneas are apparent on the first day of life. Severely affected infants require ventilatory assistance during both sleep and wakefulness. Mild to moderate developmental delay and learning disabilities are usually present in these patients. (**Aldrich 1999, p. 251**)

References

1. Aldrich M. *Sleep medicine*, 1st ed. Oxford University Press; 1999.
2. Ciftci TU, Kokturk O, Bukan N, et al. Leptin and ghrelin levels in patients with obstructive sleep apnea syndrome. *Respiration*. 2005;72(4):395–401.
3. Daube J. *Clinical neurophysiology*, 2nd ed. Oxford University Press; 2002.
4. Nishino S, Kanbayashi T. Symptomatic narcolepsy, cataplexy and hypersomnia, and their implications in the hypothalamic hypocretin (orexin) system. *Sleep MeD. Rev.* 2005;9(4):269–310.

CHAPTER 21

Sleep Studies in Clinical Practice

QUESTIONS

1. **Cataplexy is characterized by:**
   A. Appears to be related to non-rapid eye movement (NREM) sleep intrusion in waking
   B. Responds to gabapentin
   C. Can be caused by midbrain lesions
   D. Responds to imipramine

2. **Periodic limb movements of sleep (PLMS) are characterized by all except:**
   A. Usually movements of the leg, but rarely of the arms
   B. PLMS are scored only if at least two occur in sequence
   C. Movements occur for 0.5 to 5 seconds each
   D. PLMS can accompany narcolepsy or idiopathic hypersomnia

3. **Actigraphy is a useful technique to assess:**
   A. Obstructive sleep apnea
   B. Insomnia
   C. PLMS
   D. Bruxism

4. **All statements concerning multiple sleep latency test (MSLT) are correct except:**
   A. It consists of four or five 20-minute rest periods in bed in a dark room spaced 2 hours apart
   B. Sleep latency is defined as the time between lights out and beginning of stage 2 sleep

C. After sleep onset has occurred, patient is allowed to sleep for 15 minutes
D. Adequate amount of sleep must be obtained for 1 to 2 weeks before the study to ensure that the patient is not voluntarily sleep deprived

5. **All statements concerning MSLT are correct except:**
   A. If no sleep occurs during the first 20 minutes in bed, sleep latency is recorded as 20 minutes
   B. In MSLT, electroencephalogram (EEG), electrooculogram (EOG), submental electromyogram (EMG), oxygen saturation, and electrocardiogram (ECG) data are recorded
   C. For each nap, sleep latency and the occurrence of rapid eye movement (REM) sleep are noted
   D. The presence of REM in two or more naps is considered abnormal

6. **In a sleep study, the EEG and EOG suggest REM sleep, but the chin EMG shows muscle artifact. This is indicative of:**
   A. REM sleep
   B. REM behavior disorder
   C. Narcolepsy
   D. Obstructive sleep apnea

7. **In a sleep study, the EEG shows absence of alpha activity and chin EMG as well as rapid eye movements. This sleep stage is:**
   A. Awake
   B. Stage 1
   C. Stage 2
   D. REM sleep

8. **The treatment of PLMS includes all of the following except:**
   A. Imipramine
   B. Levodopa
   C. Ropinirole
   D. Pramipaxole

9. **Confusional arousals are associated with all of the following except:**
   A. They are associated with complex behaviors without full alertness
   B. They arise out of slow wave sleep
   C. They most commonly occur during the first third of the night
   D. They be associated with tachycardia, flushing, mydriasis, and sweating

10. **Sleep terrors are characterized by all except:**
    A. Episodes of agitation and fear arising abruptly from sleep during which the patient is unresponsive
    B. They tend to occur in the last third of the night
    C. They may be associated with tachycardia, flushing, mydriasis, and sweating
    D. They arise in the first third of the night

11. **All statements concerning sleep talking are correct except:**
    A. It consists of just a few words to a few sentences
    B. Ninety percent of sleep talking occurs in NREM sleep
    C. Ninety percent of sleep talking occurs in REM sleep
    D. Genetic predisposition is commonly seen

12. **Sleep paralysis is characterized by all except:**
    A. It is a common feature of narcolepsy
    B. It is a manifestation of REM intrusion into wakefulness
    C. It is a manifestation of NREM intrusion into wakefulness
    D. Patients report inability to lift even a finger during the episodes

13. **Bruxism is characterized by all except:**
    A. Grinding or clenching of the teeth during sleep
    B. It occurs at all ages
    C. Periodontal pain and tension headaches are common in these patients
    D. It occurs only in NREM sleep

14. **Alpha-delta sleep is seen in all of the following conditions except:**
    A. Schizophrenia
    B. Fibromyalgia
    C. Rheumatoid arthritis
    D. Chronic fatigue syndrome

15. **What amount of REM sleep is normal in a 6-year-old child?**
    A. Seventy percent of the total sleep
    B. Fifty percent of the total sleep
    C. Thirty percent of the total sleep
    D. Twenty-five percent of the total sleep

16. **Friedman tongue position criteria is helpful in staging which of the following?**
    A. Obstructive sleep apnea
    B. Narcolepsy
    C. Periodic leg movements of sleep
    D. Restless leg syndrome

17. **Obstructive sleep apnea is worse in:**
    A. Stage 1 sleep
    B. Stage 2 sleep
    C. Slow wave sleep
    D. REM sleep

18. **Pontine-geniculate-occipital spikes are seen in:**
    A. Stage 1 sleep
    B. Stage 2 sleep
    C. Slow wave sleep
    D. REM sleep

19. **A patient's MSLT results are as follows:**
    Mean sleep latency = 7 minutes
    REM epoch = 1 in 4 trials

    What is the diagnosis?
    A. Narcolepsy
    B. Idiopathic hypersomnolence
    C. Kleine-Levin syndrome
    D. Normal study

20. **The risk factors for obstructive sleep apnea include all except:**
    A. Hypertension
    B. Obesity
    C. Gastroesophageal reflux disease (GERD)
    D. Alcohol abuse

21. **The treatment of restless leg syndrome (RLS) includes all except:**
    A. Haloperidol
    B. Levodopa
    C. Benzodiazepines
    D. Opiates

22. **REM behavior disorder is seen in all of the following except:**
    A. Lewy body disease
    B. Parkinson disease
    C. Narcolepsy
    D. Obstructive sleep apnea

23. **The treatment of REM behavior disorder includes all of the following except:**
    A. Imipramine
    B. Levodopa
    C. Clonazepam
    D. Gabapentin

24. **REM behavior disorder can be treated by:**
    A. Clonazepam
    B. Gabapentin
    C. Pregabalin
    D. Lithium

25. **A patient with hypersomnolence has no history of cataplexy and MSLT results show:**
    Mean sleep latency = 7 minutes
    REM epoch = 3 in 4 trials

    What is the diagnosis?
    A. Narcolepsy
    B. Idiopathic hypersomnolence
    C. Kleine-Levin syndrome
    D. Monosymptomatic narcolepsy

ANSWERS

1. (D): Tricyclic antidepressants are the treatment of choice for sleep paralysis and cataplexy because they appear to act mainly through blockade of norepinephrine reuptake or serotonin reuptake and not due to anticholinergic effects. Gabapentin increases slow wave sleep but does not have any effect on cataplexy. Rarely midbrain tumors have been seen to cause cataplexy. (**Aldrich 1999, p. 171**)

2. (A): PLMS are seen in many normal, especially older sleepers with usually movements of the leg, but rarely of the arms, but may be abnormal if PLMS occur at least four in sequence with the duration of movements occurring for 0.5 to 5 seconds each. PLMS can accompany narcolepsy or idiopathic hypersomnia. PLMS >40 per hour have both a sensitivity and specificity of 81% for the diagnosis of restless leg syndrome. (**Daube 2002, p. 510**)

3. (B): Actigraphy is an inexpensive useful method for longitudinal assessment, comprising days or weeks, of sleep wake pattern. It can differentiate individuals with normal sleep pattern from those with insomnia including sleep state misperception and inadequate sleep hygiene, circadian rhythm sleep disorder. It is unable to diagnose sleep apnea, PLMS, or bruxism. (**Chokroverty 2005, p. 248**)

4. (B): MSLT is test of quantifying physiologic sleepiness during wakefulness and to determine the occurrence of REM sleep near sleep onset. It consists of four or five 20-minute rest periods in bed in a dark room spaced 2 hours apart. The sleep latency is defined as the time between lights out and beginning of any stage of sleep. After sleep onset has occurred, patient is allowed to sleep for 15 minutes to assess if REM stage occurs. It is important to confirm that adequate amount of sleep must be obtained for 1 to 2 weeks before the study to ensure that the patient is not voluntarily sleep deprived. (**Daube 2002, p. 496**)

5. (B): MSLT is test of quantifying physiologic sleepiness during wakefulness and to determine the occurrence of REM sleep near sleep onset. If no sleep occurs during the first 20 minutes in bed, sleep latency is recorded as 20 minutes. EEG, EOG, submental EMG, and ECG data are recorded. Oxygen saturation is not a part of MSLT. For each nap, sleep latency and the occurrence of REM sleep are noted. The presence of REM in two or more naps is considered abnormal. (**Daube 2002, p. 495**)

6. (B): REM sleep without atonia is scored EEG and EOG suggest REM sleep, but the chin EMG shows muscle artifact rather than the expected. This usually takes the form of a marked increase in phasic twitches but sometimes sustains tonic muscle activity. This is seen with purposeful movements, such as punching in the air during this type of sleep. This is characteristic of REM behavior disorder. (**Daube 2002, p. 500**)

7. (D): REM sleep is defined by a relatively low amplitude mixed frequency EEG similar to stage 1 sleep, in combination with markedly decreased tone in chin EMG and episodic bursts of rapid eye movements. (**Daube 2002, p. 499**)

8. (A): The most helpful treatments for PLMS are those for restless leg syndrome: levodopa and dopamine agonists, opiates, and benzodiazepines. Imipramine is not effective in PLMS. (**Aldrich 1999, p. 182**)

9. (D): Confusional arousals are characterized by a sudden arousal from sleep associated with complex behaviors without full alertness. Most commonly occur during the first third of the night in slow wave sleep. The episodes usually last just few seconds to several minutes. Patients are usually amnesic of the event. These are not associated with tachycardia, flushing, mydriasis, and sweating that is typically seen in sleep terrors. (**Aldrich 1999, p. 261**)

10. (B): Sleep terrors are episodes of agitation and fear arising abruptly from sleep during which the patient is unresponsive. These episodes are associated with high autonomic activation with tachycardia, flushing, mydriasis, and sweating. These spells usually arise within the first few hours of sleep, in the first third of the night and arise out of slow wave sleep. (**Aldrich 1999, p. 265**)

11. (C): Sleep talking consists of just a few words to a few sentences during sleep. Sleep talking is common. Approximately 90% of sleep talking occurs in NREM sleep and rest occurs in REM sleep. The cause of sleep talking is unknown. Genetic predisposition is commonly seen. (**Aldrich 1999, p. 269**)

12. (C): Sleep paralysis is a common feature of narcolepsy, however, isolated sleep paralysis is seen commonly in adolescents, occurring at least in up to 15%. It is a manifestation of REM intrusion into wakefulness. Patients often report inability to lift even a finger during the episodes. The absence of daytime sleepiness distinguishes isolated sleep paralysis from narcolepsy. (**Aldrich 1999, p. 270**)

13. (D): Sleep bruxism refers to grinding or clenching of the teeth during sleep. Bruxism occurs at all ages and affects men and women equally. It occurs during arousals at all stages of sleep including REM sleep. Periodontal pain and tension headaches are common complaints in these patients. Tooth damage can usually be prevented with a dental guard, which reduces symptoms in 80% to 90% of patients, but does not eliminate bruxing. (**Aldrich 1999, p. 279**)

14. (A): The pattern of alpha-delta sleep was once considered to be specific for fibromyalgia, but can be seen in patients with other rheumatologic disorders and chronic fatigue syndrome. It can be seen in up to 15% of asymptomatic persons. It is, however, not seen in schizophrenia. The pathophysiology for this pattern is not known. This pattern is seen with complaints of nonrestorative sleep. Treatment of sleep disturbance in patients with fibromyalgia should begin with good sleep hygiene. (**Aldrich 1999, p. 319**)

15. (D): The amount of REM sleep is approximately 50% at birth. It declines to 25% of total sleep time (the adult value) at the age of 5 years. NREM to REM cycle time increases to 80 to 90 minutes (the adult value) at the age of 5 years. In a 6-year-old child, REM sleep should comprise not >25% of the total sleep. Slow wave sleep increases to approximately 50% of total NREM sleep by approximately 1 year. Then amount of delta sleep begins to decline after the first few years of life and by the age of 9 years, delta sleep makes up approximately 22% to 28% of sleep. (**Aldrich 1999, p. 75**)

16. (**A**): Friedman tongue position (FTP) is helpful in staging the obstructive sleep apnea. This criterion is based on visualization of structures with the mouth opened widely without protruding the tongue. FTP grade 1 allows the observer to visualize the entire uvula and tonsils. Grade 2 allows visualization of the uvula but not the tonsils. Grade 3 allows visualization of the soft palate but not the uvula. Grade 4 allows visualization of the hard palate only. In general, grade 1 position has a better outcome (80% success rate) from uvulopalatopharyngoplasty. (**Friedman 2002, pp. 13–21**)

17. (**D**): Obstructive sleep apnea is much more severe during REM sleep than NREM sleep because the increased muscle atonia associated with REM sleep affects upper airway muscles and increases the likelihood of airway collapse. In addition, the apneas are longer in REM sleep because the arousal threshold is higher during REM sleep compared to NREM sleep. (**Aldrich 1999, p. 212**)

18. (**D**): A striking neurophysiologic feature of REM sleep is the occurrence of bursts of rapid firing of collections of neurons in the pons, lateral geniculate, and occipital cortex referred to as *pontine-geniculate- occipital spikes*. These spikes are frequent during REM sleep and show a rebound increase after REM sleep deprivation, appear to be generated by lateral dorsal tegmental nucleus and pedunculopontine tegmental nucleus. (**Aldrich 1999, p. 34**)

19. (**B**): Idiopathic hypersomnia refers to sleepiness despite adequate amount of sleep. The prevalence is approximately 2 to 6 in 100,000. Standard polysomnography is usually unremarkable showing normal proportions of NREM and REM sleep. An MSLT usually shows a short-sleep latency without frequent sleep-onset REM. Trials of increased sleep lead to no improvement in symptoms of daytime sleepiness or sleep latency in MSLT. (**Aldrich 1999, p. 163**)

20. (**D**): Obstructive sleep apnea (OSA) is associated with four common conditions: obesity, hypertension, GERD, and cigarette smoking. Forty percent of obese men have OSA, 25% of hypertensive patients have OSA. Sleep disordered breathing is associated with GERD and cigarette smoking. Alcohol abuse in itself is not a risk factor for OSA. (**Aldrich 1999, p. 206**)

21. (**A**): The main stay of treatment of RLS is levodopa, dopaminergic agonists, benzodiazepines, and opiates. Unfortunately, the efficacy of these medications is not always sustained and many patients require repeated dosage adjustments or medication changes. Haloperidol may not help RLS, rather may worsen the symptoms. (**Aldrich 1999, p. 179**)

22. (**D**): REM behavioral disorder is seen with degenerative disorders such as Parkinson disease, Lewy body disease, multisystem atrophy, and narcolepsy. However, sleep apnea is not associated with REM behavioral disorder. In some patients, Parkinsonism or dementia develops years after the onset of REM behavioral disorder making it first symptom of these degenerative disorders. (**Aldrich 1999, p. 273**)

23. (D): Remission in REM behavior disorder rarely occurs in cases unrelated to medications or substance abuse. Fortunately, treatment with clonazepam 0.5 to 2 mg at bedtime is effective in 80% to 90% of patients. Tolerance occasionally develops, and daytime sleepiness and memory disturbance are side effects particularly in patients with dementing illness. Imipramine and levodopa are also effective, but gabapentin is not effective. (**Aldrich 1999, p. 274**)

24. (A): Remissions rarely occur in cases unrelated to medications or substance abuse. Clonazepam 0.5 to 2 mg at bedtime is effective in 80% to 90% of patients. Clonazepam is much more effective than other benzodiazepine. Although clonazepam reduces or eliminates the pathologic behavior due to minor movements and vocalizations, incomplete atonia persists during REM sleep. If clonazepan is not effective or is not tolerated then imipramine, levodopa-carbidopa, diazepam, temazepam, clonidine, or carbamazepine are occasionally beneficial. (**Aldrich 1999, p. 274**)

25. (D): Patients with monosymptomatic narcolepsy have excessive sleepiness and an abnormal propensity to enter REM sleep prematurely, but they do not have cataplexy. Compared to patients with cataplexy they take fewer daytime naps and are less likely to have sleep paralysis, hypnogoggic hallucinations, and disturbed nighttime sleep. Human leukocyte antigen (HLA)-DR15 is more prevalent and cataplexy develops in some patients eventually. This may be a less severe manifestation, the same genetic predisposition that underlies narcolepsy–cataplexy. (**Aldrich 1999, p. 161**)

References

1. Aldrich M. *Sleep medicine*. Oxford Press; 1999.
2. Chokroverty S, Thomas R, Bhatt M. *Atlas of sleep medicine*. Elsevier Science; 2005.
3. Daube J. *Clinical neurophysiology*, 2nd ed. Oxford Press; 2002.
4. Friedman et al; *Otolaryngol Head Neck Surg* 2002;127:13–21.

SECTION 7

Autonomic Testing and Central Neurophysiology

CHAPTER 22

Autonomic Testing and Central Neurophysiology

QUESTIONS

1. **A 78-year-old man with shaking of lower extremities upon standing has:**
   A. Spinal stenosis
   B. Orthostatic tremor
   C. Cerebellar degeneration
   D. Exaggerated physiologic tremor

2. **Surface electromyogram (EMG) recording has the advantage of:**
   A. Easy accessibility of muscles
   B. Good muscle selectivity
   C. No need for a reference electrode
   D. All of the above

3. **P300 cognitive potential:**
   A. Is defined by a single generator
   B. Has a bilateral mid-parietal distribution
   C. Always needs averaging for recognition
   D. Is specific to Alzheimer disease

4. **Postganglionic sympathetic sweat glands are mediated by:**
   A. α-Adrenergic receptors
   B. β-Adrenergic receptors
   C. Nicotinic receptors
   D. Muscarinic receptors

5. **Vagal nerve effect on the heart rate is maximal at:**
   A. Early inspiration
   B. Middle of inspiration
   C. End of inspiration
   D. Heart rate is not affected with inspiration

6. **What is the normal physiologic reflex to standing?**
   A. Peripheral and splanchnic vasodilatation
   B. Peripheral and splanchnic vasoconstriction
   C. Peripheral vasoconstriction and splanchnic vasodilatation
   D. Peripheral vasodilatation and splanchnic vasoconstriction

7. **Which of the following is the most satisfactory technique for thermoregulatory sweat testing?**
   A. Hot bath
   B. Infrared lamp
   C. Incandescent lamp
   D. Environment-controlled cabinet

8. **What is the expected sweat deficit in a patient with a Pancoast tumor?**
   A. Ipsilateral hemibody loss of sweating
   B. Contralateral hemibody loss of sweating
   C. Ipsilateral head and upper trunk loss of sweating
   D. Contralateral head and upper trunk loss of sweating

9. **Dystonia is a prolonged abnormal posture caused by involuntary alternating contractions of agonist and antagonist muscles across a joint.**
   A. True
   B. False

10. **Rubral tremor typically presents with a frequency range of:**
    A. 2 to 4 Hz
    B. 4 to 8 Hz
    C. 4 to 12 Hz
    D. 14 to 18 Hz

11. **Which of the following is incorrect about the autonomic control of heart rate?**
    A. The net effect is modulation of the intrinsic firing rate of the sinus node
    B. Effect of the vagus nerve has longer duration and latency than the sympathetic effect
    C. Spontaneous fluctuations are frequent
    D. At rest, parasympathetic tone predominates

12. **Which of the following is incorrect about tremor in Parkinson's disease?**
    A. Maximal at rest and minimal with action
    B. 4 to 7 Hz frequency of alternating muscle contractions
    C. Increased tremor irregularity with advancing disease
    D. Similar tremor frequency throughout the body

13. Quantitative sudomotor axon reflex test (QSART) diagnoses C-fiber neuropathy in 80% of patients:
    A. True
    B. False

14. Upon standing, there is a surge of:
    A. Norepinephrine
    B. Acetylcholine
    C. Serotonin
    D. All of the above

15. In healthy subjects, the typical normal response to upright tilt is:
    A. Transient dizziness
    B. A transient <10% decrease in mean blood pressure
    C. A transient <20% decrease in mean blood pressure
    D. Recovery of normal blood pressure in <3 minutes

16. In anoxia-induced coma, which of the following is always predictive of no recovery?
    A. Absence of bilateral N20 with median nerve stimulation
    B. Absence of all brainstem auditory evoked potentials
    C. Absence of bilateral P100 visual evoked potentials
    D. None of the above

17. Skin vasomotor reflex can be induced by:
    A. Valsalva maneuver
    B. Inspiratory gasp
    C. Cold stimulation
    D. All of the above

18. The Valsalva maneuver test should be avoided in:
    A. Congestive heart failure
    B. Proliferative retinopathy
    C. Restrictive pericarditis
    D. All of the above

19. Denervation supersensitivity occurs with prolonged postganglionic failure:
    A. True
    B. False

20. Which of the following drugs may alter the result of the thermoregulatory sweat test?
    A. Imipramine
    B. Phenytoin
    C. Paroxetine
    D. All of the above

21. In benign paroxysmal positioning vertigo, the floating canalith is most commonly found in the:
    A. Horizontal semicircular canal
    B. Posterior semicircular canal

C. Superior semicircular canal
D. Any of the above

22. **Event-related potentials are not influenced by age:**
    A. True
    B. False

23. **Which of the following is incorrect about arterial baroreflexes?**
    A. Main effect is sympathetic innervation of splanchnic blood vessels
    B. Sympathetic outflow is mostly regulated by baroreceptors and chemoreceptors
    C. Baroreflexes induce heart rate changes in the same direction as arterial pressure
    D. Splanchnic preganglionic sympathetic neurons progressively decrease with age

24. **A normal QSART result indicates integrity of the:**
    A. Postganglionic sympathetic sudomotor axon
    B. Preganglionic sympathetic sudomotor axon
    C. Postganglionic parasympathetic sudomotor axon
    D. Preganglionic parasympathetic sudomotor axon

25. **The parasympathetic control of the heart rate is derived from the nucleus ambiguus.**
    A. True
    B. False

ANSWERS

1. (B): Orthostatic tremor or "shaky leg syndrome" occurs predominantly in the elderly upon standing causing quivering of the legs and difficulties in walking. It is diagnosed by surface EMG in the lower extremities and paraspinal muscles with the patient standing. The EMG pattern displays high amplitude 14- to 18-Hz tremor bursts, distinctive to this disorder. Exaggerated physiologic tremor is of 12 to 14 Hz frequency, best recorded in finger flexors and extensors. (**Daube 2002, p. 404**)

2. (A): Surface EMG recording lacks selectivity because of interference from other surrounding muscles. This is minimized by the use of shorter interelectrode distance with superficial muscle recording. The low frequency filter should be at least 30 Hz to minimize movement artifact. Surface EMG has the advantages of being a noninvasive method, with easy accessibility of muscles.

When evaluating deep muscles, intramuscular electrodes should be used. (**Daube 2002, pp. 400–402**)

3. (B): P300 is the most common event-related potential to be recorded by the oddball technique. It is better seen with averaging and is defined characteristically by a bilateral midparietal distribution with a latency of 300 ms and amplitude of 10 μV. Its clinical significance is still debatable and is likely to represent a complex integration of generators involved in selected attention. Its amplitude is decreased and its latency is prolonged in Alzheimers disease and other neurodegenerative disorders like Parkinsons disease or multiple sclerosis. The clinical significance of P300 potential in comatose patients is still unknown. (**Daube 2002, p. 164; Ebersole and Pedley, pp. 431–433**)

4. (**D**): In the sympathetic nervous pathway, most postganglionic neurotransmitters act through norepinephrine on adrenergic receptors. In the case of sweat glands, the sympathetic effect is mediated by acetylcholine through the M3 muscarinic receptors. (**Daube 2002, pp. 438–440**)

5. (**C**): In healthy subjects, the heart rate fluctuates with the respiratory cycle. Parasympathetic influence by the vagal nerve is highest by the end of inspiration and early expiration. This effect is minimal during early and middle inspiration. (**Daube 2002, p. 441**)

6. (**B**): Upon standing, there is pooling of venous blood to the abdomen and lower extremities. This leads to decreased venous return, mean arterial pressure, and cardiac output by approximately 20%. To maintain postural normotension, a compensatory vasoconstriction of the splanchnic, peripheral, and renal beds is mainly mediated by baroreflexes. (**Daube 2002, pp. 442–443**)

7. (**D**): The thermoregulatory sweat testing is typically used to evaluate peripheral neuropathies affecting small-diameter nerve fibers. Several techniques have been used including hot baths, infrared or incandescent lights to produce sweating. However, the most reliable and satisfactory technique is the use of controlled-environment cabinets where temperature is controlled and the whole body is equally heated. (**Daube 2002, pp. 458–462**)

8. (**C**): Pancoast tumors are tumors of the pulmonary apex, commonly compressing the sympathetic ganglion chain. The loss of sympathetic outflow causes anhidrosis or loss of sweat in the ipsilateral face, upper trunk, and arms (depending on the extent of compression of the sympathetic chain), in addition to a horner's syndrome. (**Daube 2002, pp. 462–464**)

9. (**B**): Dystonia is defined as an intense, prolonged, and simultaneous involuntary contraction of agonist and antagonist muscles producing painful stiffness across a joint and abnormal posturing. (**Daube 2002, p. 409**)

10. (**A**): Rubral tremor, also known as *Holmes tremor* typically presents with a low frequency intentional or resting tremor of 2- to 4-Hz frequency, occurring months after a midbrain insult. (**Daube 2002, pp. 402–405**)

11. (**B**): Heart rate depends on the modulation of sympathetic and parasympathetic systems on the intrinsic firing rate of the sinus node. There is frequent spontaneous autonomic fluctuation of the heart rate. At rest, the parasympathetic tone predominates over the sympathetic tone. Also, the effect of the parasympathetic tone is shorter lived than that of the sympathetic effect. (**Daube 2002, p. 441**)

12. (C): Tremor of Parkinson's disease is maximal at rest and attenuates with action. The typical tremor frequency is 4- to 7-Hz with alternating agonist and antagonist contractions. The frequency is usually similar throughout the body, becoming more regular as the disease progresses. (**Daube 2002, p. 403**)

13. (A): The QSART is abnormal in approximately 80% of patients with small fiber or C-fiber neuropathy. In this patient population, it is the most sensitive noninvasive diagnostic test. (**Daube 2002, pp. 445–447**)

14. (A): Upon standing, there is a surge of norepinephrine from postganglionic adrenergic nerve terminals. This is a compensatory mechanism to prevent hypotension caused by gravity-induced pooling of venous blood in the lower body. The plasma level of norepinephrine has been used to differentiate postganglionic from preganglionic failure. In the case of preganglionic failure, norepinephrine level is normal in the supine position and low upon standing. In the case of postganglionic failure, the supine and standing norepinephrine levels are low. (**Daube 2002, pp. 355–456**)

15. (B): Tilt table testing is typically used to diagnose dysautonomia or syncope. It can be done in different ways and be modified for individual circumstances. In healthy subjects, upright tilt can cause a transient decrement in the systolic, diastolic, and mean pressure not exceeding 10% and lasting for <1 minute. (**Daube 2002, pp. 453–454**)

16. (A): Different evoked potential modalities can be used as predictive tests of recovery in comatose patients. In the case of anoxia-induced coma, absence of bilateral cortical N20 responses with median nerve stimulation was 100% predictive of death or vegetative state within 48 hours to 1 week of coma onset. (**Ebersole and Pedley 2003, pp. 428–431**)

17. (D): Skin vasomotor response can be used to evaluate the integrity of postganglionic sympathetic adrenergic fibers. Skin blood flow is measured with a laser Doppler flowmeter or plethysmogram. Vasoconstriction can be induced by different types of maneuvers such as inspiratory gasp, standing, contralateral cold stimulation, and Valsalva maneuver. (**Daube 2002, pp. 451–452**)

18. (D): The Valsalva maneuver is a test of reflex cardiovascular response. It measures the changes in arterial pressure with sudden increase in intrathoracic and intra-abdominal pressure. This test must be avoided in patients suffering from proliferative retinopathy (risk of intraocular hemorrhage), congestive heart failure, and restrictive pericarditis (risk of angina and arrhythmias). (**Daube 2002, pp. 470–474**)

19. (A): Prolonged and widespread denervation of postganglionic sympathetic fibers may cause denervation hypersensitivity, clinically seen as exaggerated response to exogenous α-agonists. Its cellular mechanism consists of increased density, affinity, and efficacy of receptors in the sympathetic pathway. (**Daube 2002, p. 455**)

20. (A): Anticholinergic drugs including tricyclic antidepressants may inhibit sweating, causing falsely abnormal thermoregulatory sweat responses. Patients should stop such drugs at least 48 hours before the test. (**Daube 2002, pp. 464–465**)

21. **(B):** Benign paroxysmal positioning vertigo (BPPV) is an inner ear condition. Collections of calcium crystals or canaliths dislodge from the utricule and migrate into one of the semicircular canals. The posterior semicircular canal is the most commonly affected canal due to its anatomical position relative to the utricule. (**Daube 2002, pp. 417–419**)

22. **(B):** Event-related potentials (ERP) or endogenous potentials are recorded in response to external stimuli. Unlike exogenous evoked potentials (such as visual evoked potential [VEP], brainstem auditory evoked potential [BAEP], and somatosensory evoked potential [SSEP]), ERP can be recorded only in patients who can be selectively attentive to a specific stimulus. ERPs are strongly influenced by age. Latencies are prolonged in children and the elderly and shortest in mid to late teens. (**Ebersole and Pedley 2003, pp. 923–928**)

23. **(C):** Control of arterial blood pressure is mediated primarily by sympathetic innervation of blood vessels, particularly in the splanchnic bed. This sympathetic outflow is regulated by the carotid sinus and aortic baroreceptors using CN IX and CN X innervations, respectively. Baroreflexes induce changes in heart rate, in the direction opposite to the changes in arterial pressure. (**Daube 2002, pp. 441–443**)

24. **(A):** The QSART assesses the integrity of the postganglionic sympathetic sudomotor axon. Its mechanism consists of activation of sudomotor action terminals by acetylcholine or carbachol through monosynaptic transmission to the eccrine sweat glands. (**Daube 2002, pp. 448–449**)

25. **(A):** The parasympathetic control of the heart rate is derived from neurons in the nucleus ambiguus and dorsal motor nucleus of the medulla. On the other hand, the sympathetic control of the heart is derived from cervical and upper thoracic ganglia chains. (**Daube 2002, p. 441**)

References

1. Daube J. *Clinical neurophysiology*, 2nd ed. Oxford Press; 2002.
2. Ebersole J, Pedley T. *Current practice of clinical electroencephalograph*, 3rd ed. Lippincott Williams & Wilkins; 2003.

SECTION 8

Neurophysiological Intraoperative Monitoring

CHAPTER 23

Neurophysiologic Intraoperative Monitoring

QUESTIONS

1. **During left posterior tibial nerve somatosensory evoked potential (SSEP), scalp recording is performed over:**
   A. Cz and C3
   B. Cz and C4
   C. Cz and F3
   D. Cz and F4

2. **Which of the following are alarming SSEP parameters?**
   A. Loss of potentials
   B. Drop in amplitude >50%
   C. Prolongation in latency >10% or >2.5 ms
   D. All of the above

3. **During transcranial motor stimulation, the active electrode is the:**
   A. Cathode
   B. Anode
   C. Cathode or anode
   D. None of the above

4. **During SSEP monitoring, anesthesia-induced changes characteristically affect:**
   A. All four limbs equally
   B. Equally peripheral, cervical, and cortical potentials

239

C. Peripheral potentials only
D. Peripheral and cervical potentials only

5. **Electroencephalogram (EEG) slowing during intraoperative monitoring occurs when cerebral blood flow reaches:**
   A. >50 cc/100 g/minute
   B. 25 to 50 cc/100 g/minute
   C. 15 to 25 cc/100 g/minute
   D. 8 to 12 cc/100 g/minute

6. **The first sign of ischemia on EEG during intraoperative monitoring is:**
   A. Decrease in amplitude
   B. Loss of high frequency waveforms
   C. Loss of occipital rhythm
   D. All of the above

7. **Which type of electrodes is preferred in intraoperative monitoring studies?**
   A. Surface electrodes
   B. Adhesive electrodes
   C. Needle electrodes
   D. Any type of electrode

8. **Prolongation of peripheral and cortical SSEPs is caused by:**
   A. Inhalational agents
   B. Hypotension
   C. Hyponatremia
   D. Hypothermia

9. **To avoid potential complications, the use of a shunt is routinely recommended during carotid endarterectomy.**
   A. True
   B. False

10. **During vestibular schwannoma resection, electromyogram (EMG) of which muscle should be performed?**
    A. Medial rectus
    B. Orbicularis oris
    C. Masseter
    D. None of the above; brainstem auditory evoked potential (BAEP) must be performed

11. **Which of the following is true about the D wave during transcranial motor stimulation?**
    A. Consists of positive peak
    B. Is generated by the dorsal corticospinal tracts
    C. Elicited by indirect activation of the corticospinal tracts
    D. Is very sensitive to muscle relaxants

12. Which of the following is true about EMG monitoring from extraocular muscles?
    A. Performed during skull base tumor resections
    B. Needle electrodes are inserted near or in the muscles
    C. Reference electrode are placed on the contralateral side of the forehead
    D. All of the above

13. During BAEP monitoring, which of the following may cause reversible prolongation of waveform latencies?
    A. Hypothermia
    B. Hyperthermia
    C. Cerebellar retraction
    D. All of the above

14. Which of the following is true about transcranial motor evoked potentials?
    A. Magnetic stimulation is preferred
    B. Neuromuscular blockage is always indicated
    C. Stimulus interval of 1 to 3 seconds provides better waveforms
    D. Not influenced by anesthesia

15. Functional localization of the central sulcus can be performed with the use of median nerve SSEPs.
    A. True
    B. False

16. Which of the following are significant abnormal BAEP parameters?
    A. >50% Decrease in wave V amplitude
    B. >0.5 ms Increase in wave V latency
    C. Loss of wave V
    D. All of the above

17. During a T12 tumor resection, bilateral upper and lower extremity SSEPs are performed. Which of the following is true if there is progressive loss of the left cervical and cortical median potentials?
    A. Upper extremity SSEP are useless in this case
    B. Technical difficulties are the likely cause
    C. Surgeon should be quickly prompted
    D. This is a trivial finding

18. Which of the following intraoperative modalities will be most valuable in an L2 procedure?
    A. Brainstem auditory evoked potentials
    B. Free run EMG
    C. Transcranial motor potential
    D. SSEPs

19. All of the following are true about neurotonic discharges except:
    A. Often precipitated by mechanical manipulation
    B. May have various forms
    C. Damaged nerves and nerve transaction cause continuous discharges
    D. May spontaneously occur

20. **Median nerve SSEP is a valuable intraoperative test in detecting cerebral ischemia:**
    A. True
    B. False

21. **Which of the following is incorrect about pedicle screw stimulation?**
    A. Recording is performed in the corresponding muscles
    B. Thresholds below 15 mA are considered significant
    C. Bone breach causes high impedance pathway
    D. Good predictor of postoperative radiculopathy

22. **During a pituitary adenoma resection, visual evoked potentials are reliable in predicting postoperative visual loss:**
    A. True
    B. False

23. **SSEPs are conveyed by the:**
    A. Lateral lemniscus
    B. Medial lemniscus
    C. Anterior lemniscus
    D. Posterior lemniscus

24. **During intraoperative monitoring, cortical median SSEP amplitude may increase with the use of:**
    A. Isoflurane
    B. Nitrous oxide
    C. Etomidate
    D. Desflurane

25. **Which of the following is incorrect about cerebellopontine angle operations?**
    A. Cranial muscle EMG and BAEP are routinely performed
    B. Sudden loss of ipsilateral BAEP potentials is usually irreversible
    C. Preservation of facial nerve compound motor action potential (CMAP) predicts good facial nerve recovery
    D. Changes in BAEP potentials poorly correlate with postoperative hearing loss

ANSWERS

1. (A): Because of the orientation of the dipole deep inside the longitudinal fissure (for the lower extremity), the cortical potential is better recorded over the ipsilateral centroparietal region. This is referred to as the *paradoxical localization*. Conversely, in the case of the upper extremity, recording is made on the contralateral side. (**Daube 2002, p. 189; Ebersole and Pedley 2003, p. 937**)

2. (D): All of the mentioned parameters are considered to be alarming changes during SSEP monitoring. When the potentials are lost, technical causes should be ruled out first before calling it abnormal. (**Møller 2006, pp. 136–137**)

3. (B): Contrary to peripheral nerve stimulation, the active electrode in transcranial motor stimulation is the anode. Electrical stimulation is favored over magnetic stimulation when performing such studies because of more robust response and better reproducibility. (**Møller 2006, pp. 180–181**)

4. (A): Anesthesia levels may significantly alter evoked potential latencies and amplitude. Similar to hypotension and hypothermia, anesthesia-induced changes are systemic, affecting potentials of all limbs equally. Cortical potentials are the most sensitive to these changes while peripheral and cervical potentials are most resistant. (**Ebersole and Pedley 2003, p. 938; Møller 2006, p. 137**)

5. (C): The normal cerebral blood flow is 50 cc/100 g/minute. EEG may show signs of slow activity when the blood flow reaches 15 to 25 cc/100 g/minute. EEG attenuation typically occurs when blood flow reaches 12 cc/100 g/minute. (**Ebersole and Pedley 2003, pp. 949–950**)

6. (B): Loss of higher frequency waveforms is the first alarming sign of brain ischemia during carotid endarterectomy followed by voltage or amplitude asymmetry. Care must be taken to differentiate loss of higher frequency from hypocarbia, hypotension, hypothermia, and anesthesia. Of note, some inhaled anesthetic and intravenous agents such as benzodiazepines or barbiturates increase high frequency waveforms. (**Daube 2002, pp. 528–529**)

7. (C): Needle electrodes are the preferred types of electrodes used in neurophysiologic intraoperative studies. Platinum or disposable needle electrodes are preferred because they provide better stability when patient positions and manipulation is modified throughout the surgery. Taped needle electrodes are less likely to be dislodged or removed accidentally during the long surgical procedures. The disadvantages of such electrodes are the relatively higher degree of invasiveness as well as their higher impedance. (**Møller 2006, p. 41**)

8. (D): Hypothermia causes prolongation of peripheral, cervical, and cortical SSEP latencies. Inhalational agents and hypotension causes prolongation of cortical SSEP latencies. Peripheral and cervical SSEPs are more resistant to changes induced by anesthesia or hypotension. (**Møller 2006, p. 137; Ebersole and Pedley 2003, pp. 837–938**)

9. (B): During carotid endarterectomy, the risk of cerebral ischemia maybe prevented with the use of shunts to bypass the area of clamping. Routine placement of shunts carries a ten times greater risk for embolic strokes than that with the selective use of shunts. This explains the great usefulness of intraoperative EEG to prompt the use of shunts only when cerebral ischemia is detected. (**Daube 2002, p. 137**)

10. (B): Intraoperative monitoring of the facial nerve function is regarded as a valuable adjunct to vestibular schwannoma resections due to its anatomical proximity to the eighth cranial nerve. It is often performed along with intraoperative BAEP monitoring. The orbicularis oris is an easily accessible muscle innervated by the buccal branch of the facial nerve. (**Møller 2006, pp. 198–199**)

11. (B): The response from descending corticospinal tracts is recorded from the spinal cord using epidural electrodes. Transcortical stimulation of the motor pathways generates D and I waves. The D wave is assumed to be elicited by direct stimulation of the dorsal corticospinal tracts and consists of negative peaks. Similar to the I wave, it is not influenced by the use of muscle relaxants. (**Møller 2006, pp. 183–184**)

12. (D): EMG potentials from extraocular muscles are typically monitored during skull base surgeries such as carvenous sinus tumor resections. Needle electrodes are placed directly into the muscles or near the extraocular muscles. EMG activity from the lateral rectus (CN VI), inferior rectus (CN III), and superior oblique (CN IV) muscles is most commonly recorded. The reference electrode is placed on the forehead, contralateral to the monitored eye to minimize interference from the extraocular muscle contraction. (**Møller 2006, pp. 206–208**)

13. (D): During BAEP monitoring, several factors may cause prolongation of waveform latencies and attenuation of waveform amplitude. Of these are cerebellum retraction, hypothermia, or hyperthermia. These changes are transient if the cause is reversed. (**Møller 2006, pp. 105–108**)

14. (C): Transcranial magnetic and electrical motor stimulation are substantially attenuated by most inhalational anesthetics. Electrical stimulation is widely preferred over magnetic stimulation because it provides better reliability and safety. Electrical stimulation is typically delivered in a series of 1 to 3 seconds interval. This technique warrants higher amplitudes and shorter latencies of recorded potentials. (**Ebersole and Pedley 2003, pp. 940–941**)

15. (A): Localization of the central sulcus may be obtained by the use of median nerve SSEPs from cortical electrode array. The cortical N20 potential represents a horizontal dipole, positive precentrally and negative postcentrally. This phase reversal is recorded from the row of electrodes traversing the central sulcus. (**Ebersole and Pedley 2003, p. 949**)

16. (D): In general, there is consensus that a 50% or greater loss of wave V amplitude or 0.5 ms increase in wave V latency or loss of wave V are potentially abnormal alterations in the BAEP waveform. If any of these changes become permanent at the end of the procedure, there is a higher risk of postoperative hearing loss. (**Ebersole and Pedley 2003, pp. 948**)

17. (C): Intraoperative SSEP monitoring of spinal surgeries is typically performed on the four extremities. This helps in distinguishing systemic from local changes. In this case, the progressive loss of left median potentials is likely due to mechanical or ischemic changes in the left upper extremity, requiring surgeon's attention. (**Daube 2002, pp. 539–540**)

18. (B): Surgical procedures below the L1 spine level pose risk to the cauda equina rather than to the spinal cord. The absence of spinal cord below L1 makes the use of motor evoked potential insensitive. Also at the cauda equina level, there is overlap of multiple roots making the use of SSEPs insensitive as well. The most valuable test in this case would be continuous free run EMG of corresponding muscles. (**Daube 2002, pp. 551–552**)

19. (C): During intraoperative EMG, multiple muscles are monitored simultaneously. Neurotonic discharges are sensitive indicators of nerve irritation providing surgeons with feedback about potential nerve injury. Mechanical irritation such as nerve irrigation frequently produces transient neurotonic discharges. These discharges may present with varying morphology, but all consisting of motor unit potentials. Sharp nerve transaction may not produce any neurotonic discharges. Also damaged nerves are less likely to produce neurotonic discharges than intact nerves. (**Daube 2002, pp. 532–533**)

20. (A): SSEP monitoring is now commonly used as an indicator of ischemia from reduced blood flow, especially in the anterior circulation. Its value is much superior to the use of visual evoked potentials. Stimulation of the median nerve SSEP is the most widely used technique, providing more reliable waveforms. Prolongation of the central conducting time (latency interval between cervical and cortical potential) is regarded as an indicator of ischemia to the central somatosensory cortex. (**Møller 2006, pp. 139–142**)

21. (C): During transpedicular spinal fixation, postoperative radiculopathy is mainly caused by penetration of the pedicle cortex by screws. If a bone breach is present, a low impedance pathway is created, lowering the stimulation threshold. Thresholds below 15 mA are suggestive of pedicle wall breakage, often requiring repositioning of screws. (**Ebersole and Pedley 2003, pp. 945–947**)

22. (B): Visual evoked potentials have been used to monitor the visual pathways during pituitary and hypothalamic surgeries. Because of great variability of P100 latency and amplitude during operations and the poor correlation with postoperative outcomes, intraoperative visual evoked potentials are not reliable in monitoring visual pathways. (**Daube 2002, pp. 534–535**)

23. (B): SSEPs assess the ascending somatosensory pathways conveyed by the dorsal columns. Fibers from the dorsal nuclei cross the midline at the medulla to form the medial lemniscus that terminates at the somatosensory cortex. (**Møller 2006, pp. 70–71**)

24. (C): Etomidate increases amplitude of cortical SSEP while the inhaled anesthetics decrease it. The cervical and peripheral potentials are most resistant to these changes. (**Møller 2006, pp. 279–281; Ebersole and Pedley 2003, p. 938**)

25. (D): BAEP along with cranial nerve EMG monitoring are routinely performed during cerebellopontine angle procedures. Preservation of the facial nerve CMAP often predicts good facial nerve recovery. Also changes in BAEP potentials correlate well with postoperative hearing loss. Sudden loss of BAEP potentials is usually an ominous sign, representing ischemia or avulsion of the auditory nerve. (**Daube 2002, p. 536**)

References

1. Daube J. *Clinical neurophysiology*, 2nd ed. Oxford Press; 2002.
2. Ebersole J, Pedley T. *Current practice of clinical electroencephalograph*, 3rd ed. Lippincott Williams & Wilkins; 2003.
3. Møller A. *Intraoperative neurophysiological monitoring*, 2nd ed. Humana press; 2006.

INDEX

Note: Alpha numeral within brackets following page number refer to question (Q) and answer (A) number

A

Abnormal I-III interpeak latency, 123(A3)
Abnormal interpeak latency, 124(A9), 124(A12)
  of N21-P37, 123(A2)
Abnormal occipital rhythm, 70(A5)
Accessory peroneal nerve, 161(Q9), 169(A9)
Acid maltase deficiency, 181(A24)
Acid-citrase-dextrose (ACD), 183(A26)
Actigraphy, 219(Q3), 223(A3)
Action potentials, 9(Q3), 11(A3), 12(A7)
Active sleep, 54(A11)
Acute idiopathic demyelinating polyneuropathy (AIDP), 135(Q16), 141(A16)
Acute inflammatory demyelinating polyradiculoneuropathy (AIDP), 146(Q18), 146(Q21), 146(Q22), 155(A18), 156(A21), 156(A22)
Acute myopathies, 177(Q13), 181(A13)
Adult electroencephalography, 61–76
  SREDA, 61(Q1), 69(A1)
Afterhyperpolarizations, 11(Q14), 12(A14), 13(A14)
Age-related sleep changes
  in children, 213(Q18), 217(A18)
  in elderly, 213(Q17), 217(A17)
Aliasing, 16(Q6), 17(Q18), 19(A6), 19(A11), 20(A18)
Alpha coma, 68(Q25), 75(A25), 95(A42)
Alpha intrusions, 212(Q11), 217(A11)
Alpha rhythm, 74(A21)
Alpha-delta sleep, 221(Q14), 224(A14)
γ-Aminobutyric acid A (GABA) receptors, 27(Q17), 35(A17), 85(Q36), 94(A36)
Amitriptyline, 93(A31)

Amplifiers, 15(Q1), 18(A1)
Amyotrophic lateral sclerosis (ALS), 129(A1), 165(Q17), 170(A17), 183(A25), 188(Q12), 192(A12)
Analog-to-digital converter (ADC), 15(Q2), 17(Q13), 18(A2), 20(A13)
Anoxia-induced coma, 231(Q16), 234(A16)
Antidromic recording of sensory nerves, 133(Q3), 138(A3)
Antiepileptic drugs (AED), 85(Q36), 85(Q37), 88(A1), 91(A18), 94(A36), 94(A37)
Apnea, 213(Q12), 216(A12)
Apnea-hypopnea index (AHI), 218(A22)
Arterial baroreflexes, 232(Q23), 235(A23)
Attenuation, 70(A3)
Autonomic testing, 229–235
Average reference, 38(A34)
Average referential montage, 91(A16)
Axonal injury, acute, 128(Q8), 130(A8), 177(Q17), 181(A17)
Axonal regrowth, 128(Q7), 130(A7), 177(Q16), 181(A16)

B

Baclofen, 28(Q23), 36(A23)
Bancaud phenomenon, 28(Q27), 37(A27)
Barbiturate coma, SSEPs in, 121(Q13), 124(A13)
Barbiturates, 95(A43)
Benign epilepsy with centrotemporal spikes (BECTS), 48, 54(A13)
Benign epileptiform transients of sleep (BETS), 70(A4)
Benign fasciculations, 180(A10)

Benign occipital epilepsy of childhood, 47(Q18), 49(Q19), 55(A18), 55(A19)
Benign paroxysmal positioning vertigo (BPPV), 231(Q21), 235(A21)
Benign sporadic sleep spikes (BSSS), 62
Benzodiazepines, 91(A21), 95(A43)
Bilaterally delayed P100 latency, 118(Q4), 124(A4)
Biologic clock in humans, 213(Q16), 217(A16)
Bipolar montages, 26(Q11), 31(Q39), 34(A11), 39(A39), 39(A40)
Blink reflex abnormalities, 151(Q30), 157(A30), 161(Q10), 161(Q11), 169(A10), 169(A11)
Botulism, 160(Q6), 168(A6), 204(Q9), 207(A9)
Brachial plexopathy, 198(A1)
Brainstem Auditory Evoked Potential (BAEP), 103–107, 104, 235(A22)
  high frequency filter, 104(Q9), 106(A9)
  infant BAEP sweep duration, 105(Q16), 107(A16)
  low frequency filter, 104(Q8), 106(A8)
  sweep duration, 105(Q10), 106(A10)
Breach rhythm, 92(A29)
Bruxism, 221(Q13), 224(A13)
Burst suppression, 63, 71(A8)

C

Calcium spikes, 10(Q5), 12(A5)
Canavan disease, 43(Q3), 52(A3), 53(A3)
Capacitance/Capacitor, 3(Q1), 3(Q4), 5(A1), 5(A4)

247

Carbamazepine, 84(Q35), 93(A35), 94(A37)
Carotid endarterectomy, 240(Q9), 243(A9)
Carpal tunnel syndrome (CTS) electrodiagnostic studies, 134(Q7), 138(A7)
  median- versus ulnar-comparison tests, 134(Q8), 139(A8)
  proximal median neuropathy versus, 144(Q11), 154(A11)
Cataplexy, 215(A7), 219(Q1), 223(A1)
Central neurophysiology, 229–235
Cerebellopontine angle operations, 242(Q25), 245(A25)
Cerebral cortex, 10(Q10), 12(A10)
Cerebral ischemia, 242(Q20), 245(A20)
Cervical paraspinal muscles, 198(A2)
Cervical radiculopathy, 185(Q2), 189(A2)
Cervical root avulsion, 198(A2)
Chassis leakage current, 4(Q12), 6(A12)
Chassis-to-earth ground leakage current, 6(A12)
Chloral hydrate, 91(A21)
Chronic inflammatory demyelinating polyneuropathy (CIDP)
  EMG/NCS findings, 146(Q19), 146(Q23), 156(A19), 156(A23)
Clinical epilepsy, 77–95
Clonazepam, 226(A24)
Clorazepate, 89(A9)
Clozapine, 89(A9), 93(A31)
Cochlear microphonics, 104(Q7), 106(A7)
Collodion, 19(A9)
Common mode rejection, 19(A7)
Common neuromuscular disorders, EMG findings, 203–208, see also Left foot drop
  botulism, 204(Q9), 207(A9)
  congenital myasthenic syndrome, 204(Q10), 208(A10)
  congenital myasthenic syndromes, 208(A10)
  Intrinsic foot muscle atrophy, 204(Q8), 207(A8)
  peroneal neuropathy, 206(A6)
  radial neuropathy, 203(Q2), 206(A2)
Common neuromuscular disorders, nerve conduction findings, 159–172
  accessory peroneal nerve, 161(Q9), 169(A9)
  amyotrophic lateral sclerosis (ALS), 165(Q17), 170(A17)
  blink reflex, 161(Q10), 161(Q11), 169(A10), 169(A11)
  botulism, 160(Q6), 168(A6)

F-wave response, 162(Q15), 170(A15)
Lambert-Eaton myasthenic syndrome (LEMS), 160(Q4), 167(A4)
Martin-Gruber anomaly, 160(Q8), 168(A8)
in myopathic disorder, 160(Q7), 168(A7)
numbness, 161(Q12), 161(Q13), 161(Q14), 169(A12), 170(A13), 170(A14)
repetitive nerve stimulation, 159(Q2), 160(Q3), 167(A2), 167(A3)
tingling, 161(Q12), 161(Q13), 161(Q14), 169(A12), 170(A13), 170(A14)
ulnar F-wave, 164
ulnar inching, 163
Complex repetitive discharges (CRDs), 176(Q7), 178(Q23), 180(A7), 182(A23)
Concentric electromyographic (EMG) needle, 175(Q1), 179(A1)
Confusional arousals, 220(Q9), 223(A9)
Congenital blindness, 46(Q15), 55(A15)
Congenital myasthenic syndrome, 204(Q10), 208(A10)
Coumadin, 94(A37)
Creutzfeldt-Jacob disease, 78, 88(A3), 93(A31)

D
Delayed P37 absolute latency, 123(A2)
Delta brushes, 49(Q26), 56(A26), 57(A31)
Demyelination, 134(Q5), 138(A5)
Denervation supersensitivity, 231(Q19), 234(A19)
Dermatomyositis, 201(A10)
Developmental EEG characteristics, 59
Digitization, 15–21, 20(A13)
  digital EEG display, 18(A4)
  digital recording, 18(A3)
  digital signals, resolution of, 16(Q12), 19(A12)
Distinct vertex waves, 49(Q28), 57(A28)
Dorsal root ganglion (DRG), 127(Q2), 129(A2)
Drug-induced myopathies, 201(A12)
Dystonia, 230(Q9), 233(A9)

E
Ear referential montage, 89(A10)
Electric shock, 4(Q6), 6(A6)
Electrical noise, 21(A21), see also Signal-to-noise ratio
Electroconvulsive therapy (ECT), 84(Q32), 93(A32)
Electrocution risk, 4(Q11), 6(A7), 6(A11)

Electrode impedance, 16(Q10), 19(A10)
Electrodecremental response, 58(A33), 58(A34)
Electroencephalography (EEG), 10(Q10), 12(A10), 15(Q3), 18(A3), 25–41
  adult, 61–76
  alpha rhythm, 40
  beta rhythm, 40
  delta rhythm, 40
  of migraineur, 77(Q2), 88(A2)
  needle electrode, 106(A6)
  neonatal, 43–59
  pediatric, 43–59
  theta rhythm, 40
Electromyographic (EMG) findings, 17(Q20), 20(A20), 195(Q1), 198(A1), see also under Common neuromuscular disorders; Fibrillation potentials
  acute axonal injury, 177(Q17), 181(A17)
  acute myopathies, 177(Q13), 181(A13)
  axonal regrowth, 177(Q16), 181(A16)
  complex repetitive discharges (CRDs), 176(Q7), 180(A7)
  concentric electromyographic (EMG) needle, 175(Q1), 179(A1)
  end-plate noise potential, 178(Q19), 178(Q20), 178(Q21), 182(A19), 182(A20), 182(A21)
  end-plate noise, 175(Q3), 179(A3)
  fasciculations, 176(Q9), 178(Q25), 180(A9), 182(A25)
  malignant fasciculations, 176(Q10), 180(A10)
  monopolar needles, 179(Q27), 183(A27)
  myokymic responses, 178(Q26), 182(A26)
  myotonic discharges, 176(Q8), 180(A8)
  needle EMG, 179(Q29), 183(A29)
  needle EMG examination, 177(Q18), 181(A18)
  principles, 175–184
  single fiber EMG, 179(Q28), 183(A28)
  tremor EMG, 177(Q15), 181(A15)
End of chain phenomenon, 34(A11)
End-plate noise potential, 178(Q19), 178(Q20), 178(Q21), 182(A19), 182(A20), 182(A21)
Epilepsy, 38(A36)
Epileptiform discharges, 38(A36), 65(Q16), 72(A16), 80(Q9), 89(A9), 93(A31)

Epileptiform sharp transients, 30(Q36), 38(A36)
Erb's point N9, 110(Q5), 114(A5)
Erythromycin, 93(A35)
Ethosuximide, 89(A10)
Event-related potentials (ERP), 232(Q22), 235(A22)
Evoked potentials in clinical practice, 117–124
  abnormal I-III interpeak latency, 117(Q3), 123(A3)
  abnormal interpeak latency of N21-P37, 120(Q9), 124(A9)
  abnormal P100 latency, 117(Q1), 123(A1)
  barbiturate coma, 121(Q13), 124(A13)
  bilaterally delayed P100 latency, 118(Q4), 123(A4)
  brain death and, 118(Q6), 123(A6)
  N13-N20 latency, 119(Q12), 124(A12)
  P37 latency, 117(Q2), 123(A2)
  P37 latency, 119(Q11), 124(A11)
Excitatory postsynaptic potential (EPSP), 10(Q9), 12(A9)
Extensor digitorum brevis (EDB) muscle, 169(A9)
Eye flutter, 95(A42)
Eye movements, 36(A26)
Eyelid flutter, 94(A42)

F
Fasciculations, 176(Q9), 178(Q25), 180(A9), 180(A10), 182(A25)
  benign, 180(A10)
  malignant, 176(Q10), 180(A10)
Fibrillation potentials, 176(Q4), 176(Q5), 179(A4), 180(A5)
Flutter, 87
Focal cortical lesions, 61(Q3), 70(A3)
Focal epileptiform discharge, 6(Q9), 33(A9)
Focal motor status epilepticus, 52(A4)
Focal polymorphic delta activity, 94(A41)
Focal slow waves, 29(Q31), 30(Q33), 37(A31), 38(A33)
Foramen ovale electrodes, 70(A6)
Friedman tongue position (FTP), 221(Q16), 225(A16)
Frontal intermittent rhythmic delta activity (FIRDA), 37(A30), 79, 89(A8)
Frontal lobe complex partial seizures, 79(Q12), 90(A12)
Frontal lobe epilepsy, 90(A12)
Frontal sharp transients, 49(Q27), 57(A27)
Frontotemporal electrodes, 35(A20)
F-wave, 143(Q1), 152(A1)
F-wave latency, 151(Q28), 153(A1), 153(A2), 153(A3), 157(A28), 162(Q15), 170(A15)

G
Gabapentin, 223(A1)
Generalized paroxysmal fast activity (GPFA), 67(Q27), 68, 73(A22), 74(A27)
Generalized periodic epileptiform discharge (GPED), 86, 94(A40)
Glossokinetic artifact, 37(A29)
Gluteus maximus muscle, 186(Q6), 191(A6)
Ground loop, 36(A21)

H
H reflex, 143(Q4), 153(A4)
Hearing threshold, 106(A4)
Hemispheric voltage attenuation, 56(A21)
Hereditary motor sensory neuropathy-type 1 (HMSN-1), 135(Q14), 135(Q15), 140(A14), 141(A15)
High-pass filter, 16(Q8), 17(Q15), 19(A8), 20(A15)
Hippocampal ictal onset, 67(Q29), 74(A29)
Hodgkin-Huxley model, 11(A3)
Holmes tremor, 233(A10)
Human leukocyte antigen (HLA)-DR15, 226(A25)
Hyperacute injury, 130(A8), 181(A17)
Hyperammonemia, 90(A14)
Hyperkalemic periodic paralysis, 181(A24)
Hypermagnesemia, 89(A11)
Hypersomnolence, 222(Q25), 226(A25)
Hyperventilation, 38(A35), 55(A17), 57(A29), 77(Q4), 88(A4)
  induced myokyma, 183(A26)
Hypnagogic hypersynchrony, 44(Q7), 53(A7)
Hypnic jerks, 64(Q17), 72(A17)
Hypnogoggic hallucinations, 216(A7)
Hypnopompic hallucinations, 216(A7)
Hypnopompic hypersynchrony, 44(Q7), 53(A7)
Hypocretin/orexin deficiency, 214(Q21), 217(A21)
Hypoglycemia, 88(A4)
Hypopnea, 213(Q12), 216(A12)
Hypothermia, 243(A8)
Hyps, 81

I
Ictal discharges, 94(A42)
Ictal EEG, 83(Q27), 92(A27)
Idiopathic hypersomnia, 225(A19)
Imipramine, 93(A31)
Impedance, 16(Q10), 17(Q19), 19(A10), 20(A19)
Inching techniques in ulnar nerve evaluation, 149(Q27), 157(A27)
Inclusion body myositis (IBM), 201(A11)
Inductor (coil), 4(Q5), 5(A5)
Infantile spasms, 50(Q33), 57(A33)
Inferior frontal or anterior temporal electrodes, 35(A20)
Inherited demyelinating polyneuropathies, 140(A14)
Inhibitory postsynaptic potentials (IPSPs), 10(Q6), 12(A6)
Insomnia, 214(Q24), 218(A24)
Instrumentation, 15–21
Interictal epileptiform discharges, 84(Q31), 93(A31)
Intraoperative electrocorticography, 64(Q16), 72(A16)
Intraoperative monitoring, 239–245
  carotid endarterectomy, 240(Q9), 243(A9)
  cerebellopontine angle operations, 242(Q25), 245(A25)
  cerebral ischemia, 242(Q20), 245(A20)
  EEG, 240(Q5), 243(A5)
  electrodes in, 240(Q7), 243(A7)
  extraocular muscles, 241(Q12), 244(A12)
  pedicle screw stimulation, 242(Q21), 245(A21)
  pituitary adenoma resection, 242(Q22), 245(A22)
  transcranial motor stimulation, 240(Q11), 244(A11)
  vestibular schwannoma resection, 240(Q10), 243(A10)
Intrinsic foot muscle atrophy, 204(Q8), 207(A8)
Ion movements, 11(Q12), 12(A12)
Ipsilateral ear montage, 26(Q10), 33(A10)
Ischemia, 240(Q6), 243(A6)
Ischemic myelopathy, 123(A5)

J
Juvenile myoclonic epilepsy, 80(Q15), 90(A15)

K
K-complex, 211(Q3), 215(A3)
Kirchhoff's law, 3(Q3), 5(A3)

L

Lambda waves, 65(Q18), 75(A18), 74(A18)
Lambert-Eaton myasthenic syndrome (LEMS), 160(Q4), 167(A4)
Lamotrigine, 94(A39)
Landau-Kleffner syndrome, 52(A5)
Lateral hypothalamus (LHA), 217(A21)
Leakage current, 5(Q13), 6(A13)
   chassis-to-earth ground leakage current, 6(A12)
Left abductor policis brevis muscles, 202(A14)
Left biceps muscle, 202(A14)
Left deltoid muscle, 187(Q9), 191(A9), 202(A14)
Left first dorsal interosseous muscle, 187(Q9), 191(A9), 202(A14)
Left foot drop
   from a common peroneal neuropathy at the knee, 204(Q4), 206(A4)
   from a common peroneal neuropathy in the thigh, 204(Q6), 206(A6)
   with peroneal neuropathy at the fibular head, 204(Q5), 206(A5)
Left pronator teres muscle, 187(Q9), 191(A9), 202(A14)
Left tricep muscle, 187(Q9), 191(A9), 202(A14)
Lennox-Gastaut syndrome (LGS), 54, 74(A27)
Leptin, 214(Q22), 218(A22)
Lethal shock, 4(Q9), 6(A9)
Levetiracetam, 94(A37)
Linked ear montages, 39(A39)
Lithium, 93(A31)
Longitudinal bipolar montage, 39(A37), 72(A15), 72(A20), 88(A3), 94(A42)
Low-pass filter, 17(Q14), 17(Q16), 20(A14), 20(A15), 20(A16)
Lumbosacral paraspinal muscles, 186(Q6), 191(A6)

M

Malignant fasciculations, 180(A10)
Martin-Gruber anastomosis (MGA), 141(A20), 168(A8)
Martin-Gruber anomaly, 160(Q8), 168(A8)
Masking noise, 104(Q5), 106(A5)
Maximal allowed leakage current, in an ICU patient, 4(Q7), 6(A7)
Medial gastrocnemius muscle, 186(Q6), 191(A6)
Median versus ulnar comparison tests, 147, 152
   CTS, 134(Q8), 139(A8), 144(Q9), 154(A9)
Median versus ulnar sensory nerve conduction (SNC), 147
Mesial temporal lobe epilepsy (MTLE), 89(A5)
Middle trunk lesions, 198(A3)
Migraine, 88(A2)
Migraineur, 77(Q2), 88(A2)
Mitral valve replacement surgery, 151(Q29), 157(A29)
Mononeuropathy multiplex pattern, 135(Q13), 139(A13)
Mononeuropathy multiplex, 145(Q17), 155(A17)
Monopolar needles, 179(Q27), 183(A27)
Monosymptomatic narcolepsy, 226(A25)
Motor nerve conduction studies, 143–158
   AIDP, 146(Q18), 146(Q21), 146(Q22), 155(A18), 156(A21), 156(A22)
   blink reflex abnormalities, 151(Q30), 157(A30)
   chronic inflammatory demyelinating polyneuropathy (CIDP), 146(Q19), 146(Q23), 155(A19), 156(A23)
   F-wave, 143(Q1), 153(A1)
   H reflex, 143(Q4), 153(A4)
   inching techniques, 150(Q27), 157(A27)
   median versus ulnar comparison tests, 144(Q9), 154(A9)
   mononeuropathy multiplex, 145(Q17), 155(A17)
   multifocal motor neuropathy, 146(Q20), 156(A20)
   myasthenia gravis, 144(Q6), 154(A6)
   normal F-wave latency, 143(Q2), 153(A2)
   radial neuropathy, 145(Q16), 155(A16)
   repetitive nerve stimulation (RNS), 144(Q5), 153(A5)
   routine nerve conduction studies (NCSs), 144(Q8), 154(A8)
   routine ulnar NCS, 145(Q13), 145(Q14), 154(A13), 155(A14)
   single fiber electromyographic (SFEMG) tracings, 147(Q25), 157(A25)
Motor neuron disease, 185–193
Motor unit action potential (MUAP), 20(A20), 201(A11)
Motor vehicle accident (MVA), 195(Q1), 198(A1)
MUAP, 201(A11)
Multifocal motor neuropathy, 146(Q20), 156(A20)
Multiple sleep latency test (MSLT), 212(Q10), 217(A10), 219(Q4), 220(Q5), 223(A4), 223(A5)
Myasthenia gravis, 144(Q6), 154(A6), 159(Q1), 167(A1), 191(A9)
Myelin sheath, 10(Q8), 12(A8)
Myelinated fibers, 133(Q4), 138(A4)
Myelination, 127(Q3), 129(A3)
Myoclonic jerks, 72(A17), 81(Q17), 91(A17)
Myoclonic seizures, 90(A15)
Myokymia, 183(A26)
Myopathic disorder, 178(Q26), 182(A26), 183(A26), 195–202
   nerve conduction studies, 129(Q11), 131(A11), 160(Q7), 168(A7)
Myotonic discharges, 176(Q8), 180(A8)

N

N11–13 complex, 110(Q6), 114(A6)
N13-N20 latency, 120(Q12), 124(A12)
N21-P37 latency, 120(Q9), 124(A9)
Narcolepsy, 212(Q7), 212(Q9), 215(A7), 216(A8), 216(A9), 217(A9), 226(A25)
Nascent motor units, 181(A14)
Needle electrodes, 106(A6)
Needle electromyography (EMG), 165(Q22), 171(A22), 177(Q18), 179(Q29), 181(A18), 183(A29), 195(Q2), 196(Q3), 196(Q4), 198(A2), 198(A3), 198(A4)
Neocortex, 10(Q11), 12(A11)
Neonatal electroencephalography, 43–59, 43(Q1), 51(A1)
   electrode placement for, 44
   length of, 45(Q10), 53(A10)
Neonatal myasthenia, 167(A1)
Neonatal seizures, 48(Q23), 56(A23)
Nernst potential, 9(Q4), 11(A4)
Nerve biopsy, 123(A5)
Nerve conduction studies, 127–131, see also Common neuromuscular disorders; Motor nerve conduction studies; Sensory nerve conduction studies
   acute axonal injury, 128(Q8), 130(A8)
   axonal regrowth, 128(Q7), 130(A7)
   dorsal root ganglion (DRG), 127(Q2), 129(A2)
   myelination, 127(Q3), 129(A3)
   in myopathic disorder, 129(Q11), 131(A11)
   routine sural nerve sensory study, 127(Q1), 129(A1)
Net ionic flux, 12(A12)

Neuralgic amyotrophy, 199(A6)
Neuromuscular junction (NMJ), 181(A18)
Neuronal membrane depolarization, 10(Q7), 12(A7)
Neurophysiologic signals, 18(Q22), 21(A22)
Nonepileptic psychogenic seizures, 83(Q28), 92(A28)
Nonepileptiform sharp transients, 30(Q36), 38(A36)
Nonpolarizable electrodes, 25(Q1), 32(A1)
Non–rapid eye movement (NREM) sleep, 52(A2), 90(A13)
Normal F-wave latency, 143(Q2), 153(A2)
Numbness, 161(Q12), 161(Q13), 161(Q14), 169(A12), 170(A13), 170(A14)

O

Obstructive sleep apnea (OSA), 212(Q5), 215(A5), 221(Q17), 222(Q20), 225(A17), 225(A20)
Occipital abnormalities, 55(A15)
Occipital intermittent rhythmic delta activity (OIRDA), 77(A30)
Ohm's law, 6(A10)
Olfactory hallucination, 78(Q5), 88(A5)
Ondine's curse, 214(Q25), 218(A25)
Orbicularis, 244(A10)
Orthostatic tremor, 232(A1)
Oxcarbazepine, 93(A36), 94(A39)

P

P100 latency, 99(Q4), 100(Q5), 101(A4), 101(A5)
  abnormal, 117(Q1), 123(A1)
    bilaterally delayed, 118(Q4), 123(A4)
  interocular P100 latency, 100(Q10), 101(A10)
P300 cognitive potential, 229(Q3), 232(A3)
P37 latency, 120(Q11), 124(A11)
Pancoast tumors, 230(Q8), 233(A8)
Paradoxical localization, 242(A1)
Paralysis, sleep, 221(Q12), 224(A12)
Parkinson disease, 230(Q12), 234(A12)
Paroxysmal depolarization shift (PDS), 12(A5), 33(A9)
Paroxysmal response, 63(Q13), 71(A13)
Partial epilepsy, 83(Q26), 90(A12), 92(A26)

Pattern shift visual evoked potential (PSVEP), 99(Q1), 99(Q3), 100(A1), 101(A3)
  signal-to-noise ratio for, 101(A3)
Pediatric electroencephalography, 43–59
Pedicle screw stimulation, 242(Q21), 245(A21)
Penicillamine, 167(A1)
Periodic lateralized epileptiform discharges (PLEDs), 37(A30), 67(Q26), 73(A26)
Periodic limb movements of sleep (PLMS), 213(Q15), 217(A15), 219(Q2), 220(Q8), 223(A2), 223(A8)
Periodontal pain, 224(A13)
Peroneal neuropathy, 206(A6)
Phenytoin, 88(A1), 93(A37)
  discontinuation, 77(Q1), 88(A1)
Photic stimulation, 30(Q35), 38(A35)
Photomyoclonic response, 63(Q10), 71(A10)
Photoparoxysmal or photoconvulsive response, 71(A13)
Photosensitive epilepsies, 93(A34)
Pituitary adenoma resection, 242(Q22), 245(A22)
Plexopathies, 195–202
Polymyositis, 201(A10)
Pontine-geniculate-occipital spikes, 221(Q18), 225(A18)
Positive occipital sharp transients (POSTs), 55(A15)
Positive waves, 178(Q22), 182(A22)
Postganglionic sympathetic sweat glands, 229(Q4), 233(A4)
Postsynaptic potentials, 12(A6)
Postsynaptic potentials, 36(A24)
Proximal median neuropathy, CTS versus, 144(Q11), 154(A11)
Proximal myopathy, 201(A12)
Pyramidal cells, 12(A10)

Q

Quantitative sudomotor axon reflex test (QSART), 231(Q13), 234(A13)
Quantization, 20(A13)
Quiet sleep, 49(Q31), 54(A11), 57(A31), 58(A32)

R

Radial neuropathy, 145(Q16), 155(A16), 203(Q2), 206(A2)
  denervation pattern, 203(Q3), 206(A3)
Radiculopathies, 185–193
Raphe magnus, 217(A16), 217(A19)

Rapid eye movement (REM) sleep, 211(Q1), 211(Q4), 212(Q6), 214(A1), 215(A4), 215(A6)
Rarefaction, 106(A7), 106(A11)
Rasmussen syndrome, 44(Q4), 52(A4)
Reference, 38(A34), 39(A39)
Reflex epilepsy, 84(Q34), 93(A34)
Repetitive nerve stimulation (RNS), 144(Q5), 153(A5)
  decrement on, 159(Q2), 160(Q3), 167(A2), 167(A3)
Resistance/Resistor, 4(Q8), 6(A8)
Resolution of digital signals, 16(Q12), 19(A12)
Resting membrane potential of neurons, 9(Q1), 11(A1), 11(A2)
Restless leg syndrome (RLS), 214(Q20), 217(A20), 222(Q21), 225(A21)
Reversal of polarity, 39(A37)
Reye syndrome, 44(Q6), 53(A6)
Rhomboids, 195(Q1), 198(A1), 198(A2)
Rhythmic activity, 37(A30)
Rhythmic midtemporal discharges (RMTD), 65, 72(A15)
Rhythmic temporal theta burst of drowsiness (RTTBD), 73(A15)
Right frontal epilepsy, 84(Q35), 93(A35)
Ripples of prematurity, 58(A36)
Routine sural nerve sensory study, 127(Q1), 129(A1)
Rubral tremor, 230(Q10), 233(A10)
Rudimentary vertex waves, 57(A28)
Rules of polarity, 41

S

Saccadic eye movements, 72(A18)
Sampling Theorem, 19(A6), 20(A13)
Scalp-recorded electroencephalogram (EEG), 26(Q5), 26(Q8), 27(Q14), 27(Q15), 28(Q24), 28(Q25), 32(A5), 32(A6), 33(A8), 35(A14), 35(A15), 36(A24), 36(A25)
Seizures, 79(Q11), 88(A5), 89(A11), 239(Q1), 242(A1)
  frontal lobe complex partial seizures, 79(Q12), 90(A12)
  of reflex epilepsy, 93(A34)
Sensory nerve action potential (SNAP), 20(A17), 196(Q4), 198(A4)
Sensory nerve conduction studies, 133–142
  acute idiopathic demyelinating polyneuropathy (AIDP), 135(Q16), 141(A16)
  antidromic recording of sensory nerves, 133(Q3), 138(A3)
  Carpal tunnel syndrome, 134(Q7), 138(A7)
  demyelination, 134(Q5), 138(A5)

Sensory nerve conduction studies (*continued.*)
  fiber peripheral neuropathy, 135(Q12), 139(A12)
  gain, 133(Q1), 138(A1)
  hereditary motor sensory neuropathy-type 1 (HMSN-1), 135(Q14), 135(Q15), 140(A14), 141(A15)
  median- versus ulnar-comparison tests, 134(Q8), 139(A8)
  mononeuropathy multiplex pattern, 135(Q13), 139(A13)
  myelinated fibers, 133(Q4), 138(A4)
  patients with radiculopathy, 136(Q17), 141(A17)
  routine nerve conduction studies, 134(Q6), 138(A6)
Shaky leg syndrome, 232(A1)
Sharp wave, 27(Q19), 35(A19), 38(A36)
Signal-to-noise ratio (SNR), 17(Q17), 20(A17)
Single fiber electromyographic (SFEMG) tracings, 147(Q25), 157(A25)
Single fiber EMG, 179(Q28), 183(A28)
Skin vasomotor reflex, 231(Q17), 234(A17)
Skull defect, 92(A29)
Sleep/Sleep medicine, 211–218, 219–226, *see also* Rapid eye movement (REM) sleep
  actigraphy, 219(Q3), 223(A3)
  active sleep, 54(A11)
  age-related sleep changes, 213(Q17), 213(Q18), 217(A17), 217(A18)
  alpha intrusions, 212(Q11), 217(A11)
  alpha-delta sleep, 221(Q14), 224(A14)
  apnea, 213(Q12), 216(A12)
  asynchronous sleep spindles, 53(A8)
  biologic clock in humans, 213(Q16), 217(A16)
  Bruxism, 221(Q13), 224(A13)
  cataplexy, 219(Q1), 223(A1)
  cerebrospinal fluid (CSF) hypocretin-1 measures, 217(A21)
  confusional arousals, 220(Q9), 223(A9)
  Friedman tongue position criteria, 221(Q16), 225(A16)
  hypersomnolence, 222(Q25), 226(A25)
  hypocretin/orexin deficiency, 214(Q21), 217(A21)
  hypopnea, 213(Q12), 216(A12)
  insomnia, 214(Q24), 218(A24)
  K-complex, 211(Q3), 215(A3)
  leptin, 214(Q22), 218(A22)
  multiple sleep latency test (MSLT), 212(Q10), 217(A10) 219(Q4), 220(Q5), 223(A4), 223(A5)
  narcolepsy, 212(Q7), 215(A7)
  nucleus role in, 214(Q19), 217(A19)
  obstructive sleep apnea, 212(Q5), 215(A5), 221(Q17), 225(A17)
  Ondine's curse, 214(Q25), 218(A25)
  paralysis, 221(Q12), 224(A12)
  periodic limb movements of sleep (PLMS), 213(Q15), 217(A15), 219(Q2), 220(Q8), 223(A2), 223(A8)s
  pontine-geniculate-occipital spikes, 221(Q18), 225(A18)
  quiet sleep, 54(A11)
  REM behavior disorder, 222(Q22), 225(A22)
  restless leg syndrome (RLS), 214(Q20), 217(A20), 222(Q21), 225(A21)
  sleep hygiene, 214(Q23), 218(A23)
  spindles, 13(A13), 43(Q2), 51(A2)
  stage 1 sleep, 213(Q13), 216(A13)
  stage 2 sleep, 211(Q2), 215(A2)
  stage slow wave sleep, 213(Q14), 216(A14)
  stage, 50(Q32), 57(A32)
  talking, 221(Q11), 224(A11)
  terrors, 220(Q10), 223(A10)
  wakefulness test, 212(Q8), 216(A8)
Slow spike discharge, 54(A14)
Small sharp spikes (SSS), 70(A4)
Somatosensory evoked potentials (SEP), 20(A17), 109–116, 239(Q1), 242(A1)
  better resolution, procedures for, 111(Q16), 115(A16)
  bilateral stimulation, 111(Q13), 111(Q14), 115(A13), 115(A14)
  in newborns, 113(Q24), 116(A24)
  N9 of, 114(A6)
  patient's height recording, 112(Q18), 115(A18)
  recording, 109(Q1), 109(Q2), 109(Q3), 113(A1), 113(A2), 113(Q3)
  signal-to-noise ratio, 111(Q15), 115(A15)
  spinal cord values, 112(Q19), 115(A19)
  SSEP test, problem encountered with, 111(Q10), 115(A10)
  stimulus intensity, 111(Q11), 111(Q12), 115(A11), 115(A12)
Sphenoidal electrodes, 29(Q28), 37(A28)
Spike discharge, 78(Q7), 82(Q19), 89(A7), 91(A19)
Spikes, 38(A36)
Spindle coma, 91(A24), 94(A42)
Spindle duration, 52(A2)
Spontaneous activity, 184(A29), 200(A9)
  in needle EMG, 183
Squeak phenomenon, 74(A28)
Stage 1 sleep, 213(Q13), 216(A13)
Stage 2 sleep, 211(Q2), 215(A2)
Stage slow wave sleep, 213(Q14), 216(A14)
Steroid myopathy, 201(A12)
Subacute sclerosing panencephalitis (SSPE), 54(A12)
Subclinical rhythmic electrographic discharge of adult (SREDA), 61(Q1), 69(A1)
Subdural grids, 70(A6)
Suprascapular entrapment, 199(A6)
Suprascapular nerve, 200
Suprascapular notch, 200(A6)
Surface electromyogram (EMG), 229(Q2), 232(A2)
Syncopal episode, 65(Q19), 72(A19)
Syncope, 72(A19)

T
Talking, sleep, 221(Q11), 224(A11)
Taped needle electrodes, 243(A7)
Tarsal tunnel syndrome, 129(A1)
Temporal intermittent rhythmic delta activity (TIRDA), 37(A30), 70, 74(A30)
Temporal lobe epilepsy, 85(Q37), 93(A37)
Temporal spikes, 63(Q12), 71(A12)
Tension headaches, 224(A13)
Tensor fascia lata muscle, 186(Q6), 191(A6)
Terrors, sleep, 220(Q10), 223(A10)
Tetrodotoxin (TTX), 28(Q22), 36(A22)
Thalamus, 11(Q13), 12(A13)
Thermoregulatory sweat test, 230(Q7), 231(Q20), 233(A7), 234(A20)
Thoracic outlet syndrome (TOS), 199(A5)
Tiagabine, 93(A36)
Tibial nerve, 140
Tibialis anterior muscle, 186(Q6), 191(A6)
Tilt table testing, 234(A15)
Time constant (TC) of low-frequency filters, 17(Q16), 20(A16)
Tingling, 161(Q12), 161(Q13), 161(Q14), 169(A12), 170(A13), 170(A14)

Topiramate, 93(A36), 93(A37), 94(A39)
Tracé alternant, 57(A31)
Tracé discontinu, 54(A11), 56(A20)
Transcranial motor stimulation, 239(Q3), 240(Q11), 241(Q14), 243(A3), 244(A11), 244(A14)
Trapezius, 199(A6)
Tremor EMG, 177(Q15), 181(A15)
Tricyclic antidepressants, 223(A1)
Triphasic waves, 90(A14)

U

Ulnar F-wave, 164
Ulnar inching, 150, 163
Ulnar motor nerve conduction (MNC) demyelinating neuropathy, 150
Unilateral P100 delay, 123(A8)
Upper trunk lesions, 198(A2)

V

Vagal nerve effect, 230(Q5), 233(A5)
Vagal nerve stimulator (VNS), 83(Q25), 92(A25)
Valsalva maneuver test, 231(Q18), 234(A17), 234(A18)
Vastus medialis muscle, 186(Q6), 191(A6)
Ventricular fibrillation, 3(Q2), 5(A2)
Vertex reference, 38(A34)
Vestibular schwannoma resection, 240(Q10), 243(A10)
Video EEG monitoring, 83(Q28), 92(A28)
Vigabatrin, 94(A36)
Visual acuity, 100(Q6), 101(A6)
Visual evoked potential (VEP), 99–102, 119(Q8), 123(A8)

W

Wakefulness, 56(A25), 216(A8)
  EEG rhythm during, 49(Q30), 57(A30)
  test, 212(Q8), 216(A8)
Wallerian degeneration, 181(A17)
Wave discharges, 54(A14), 78(Q7), 82(Q19), 89(A7), 91(A19)
Wave I, 103(Q1), 104(Q7), 105(Q12), 105(Q13), 105(Q14), 106(A1), 106(A7), 107(A12), 107(A13), 107(A14)
Wave III, 105(Q17), 107(A17)
Wave repetition, 37(A30)
Wave V, 103(Q3), 106(A3)
West syndrome, 45(Q9), 53(A9)

Z

Zygomatic electrodes, 29(Q28), 37(A28)